Karl K. Sabelfeld and Nikolai A. Simonov

Stochastic Methods for Boundary Value Problems

Also of Interest

Random Fields and Stochastic Lagrangian Models: Analysis and Applications in Turbulence and Porous Media
Karl K. Sabelfeld, 2012
ISBN 978-3-11-029664-8, e-ISBN 978-3-11-029681-5

Spherical and Plane Integral Operators for PDEs: Construction, Analysis, and Applications
Karl K. Sabelfeld, Irina A. Shalimova, 2013
ISBN 978-3-11-031529-5, e-ISBN 978-3-11-031533-2

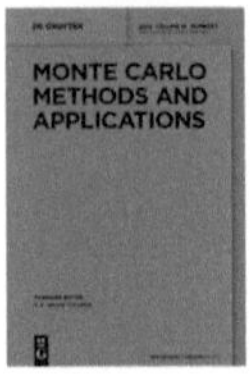

Monte Carlo Methods and Applications
Karl K. Sabelfeld (Managing Editor)
ISSN 0929-9629, e-ISSN 1569-3961

Computational Methods in Applied Mathematics
Carsten Carstensen (Editor-in-Chief)
ISSN 1609-4840, e-ISSN 1609-9389

Radon Series on Computational and Applied Mathematics
Ulrich Langer, Hansjörg Albrecher, Heinz W. Engl, Ronald H. W. Hoppe, Karl Kunisch, Harald Niederreiter, Christian Schmeiser (Ed.)
ISSN 1865-3707

Karl K. Sabelfeld and Nikolai A. Simonov

Stochastic Methods for Boundary Value Problems

Numerics for High-Dimensional PDEs and Applications

DE GRUYTER

Authors
Prof. Dr. Karl K. Sabelfeld
Novosibirsk University (NSU)
Russian Academy of Sciences
Siberian Branch
Institute of Computational Mathematics and Mathematical Geophysics
Lavrentjeva 6
630090 Novosibirsk
RUSSIA
sabelfeld.karl@yahoo.de

Nikolai A. Simonov
Russian Academy of Sciences
Siberian Branch
Institute of Computational Mathematics and Mathematical Geophysics
Lavrentjeva 6
630090 Novosibirsk
RUSSIA
nasfsu@gmail.com

ISBN 978-3-11-047906-5
e-ISBN (PDF) 978-3-11-047945-4
e-ISBN (EPUB) 978-3-11-047916-4
Set-ISBN 978-3-11-047946-1

Library of Congress Cataloging-in-Publication Data
A CIP catalog record for this book has been applied for at the Library of Congress.

Bibliographic information published by the Deutsche Nationalbibliothek
The Deutsche Nationalbibliothek lists this publication in the Deutsche Nationalbibliografie; detailed bibliographic data are available on the Internet at http://dnb.dnb.de.

Cover image: Rui Tho/Hemera/thinkstock
Typesetting: Konvertus, Haarlem
Printing and binding: CPI books GmbH, Leck
♾ Printed on acid-free paper
Printed in Germany

www.degruyter.com

Preface

The monograph is devoted to random walk-based stochastic algorithms for solving high-dimensional boundary value problems of mathematical physics and chemistry. The book includes stochastic algorithms of two classes of random walk methods: (1) methods where the random walks live on the boundary and (2) methods where the random walks are Markov chains of random points inside the domain. The random walk-on-boundary methods are presented in Chapters 1–6 for the main boundary value problems of the heat, electrostatics and elastostatics potential theory.

The methods based on random walks inside the domain are intrinsically related with the probabilistic representations of solutions to PDEs in the form of Wiener path integrals. Therefore, different ideas from probabilistic representations, such as the first passage time, exit point and survival probability concepts, are used to construct highly efficient stochastic algorithms. We show how these methods can be used to solve some applied problems, in particular the simulation of kinetics of electron-hole recombination in semiconductors, cathodoluminescence imaging method, capacitance calculations, evaluation of electrostatic energy for macromolecules in salt solutions and others. The stochastic algorithms based on random walks are grid free; they are all scalable and can be easily parallelized. Along with the probabilistic representations, we use also a functional approach based on Green functions and local integral equations, which are equivalent to the original boundary value problems.

The book is written for mathematicians who work in the field of partial differential and integral equations, for physicists and engineers dealing with computational methods and applied probability, and for students and postgraduates studying mathematical physics and numerical mathematics.

The support of Russian Science Foundation under grant No. 14-11-00083 is kindly acknowledged.

Contents

1 Introduction

It is well known that the random walk methods for boundary value problems (BVPs) for high-dimensional domains with complex boundaries are quite efficient, especially if it is necessary to find the solution not in all the points of a grid but only in some marked points of interest. One of the most impressive features of the Monte Carlo methods is the possibility to calculate probabilistic characteristics of the solutions to BVPs with random parameters (random boundary functions, random sources and even random boundaries).

Monte Carlo methods for solving partial differential equations (PDEs) are based on the following:

(a) classical probabilistic representations in the form of Wiener or diffusion path integrals,
(b) probabilistic interpretation of integral equations equivalent to the original BVP, which results in representations of the solutions as expectations over Markov chains.

In approach (a), diffusion processes generated by the relevant differential operator are simulated using numerical methods for solving ordinary stochastic differential equations [29]. To achieve the desired accuracy, it is necessary to take the discretization step small enough, which results in long random trajectories simulated.

For PDEs with constant coefficients, however, it is possible to use the strong Markov property of the Wiener process and create much more efficient algorithms first constructed for the Laplace equation [76] known as the walk-on-spheres (WOS) method.

This algorithm was later justified in the framework of approach (b) by passing to an equivalent integral equation of the second kind with generalized kernel and using the Markov chain that 'solves' this equation. This approach was developed for general second-order scalar elliptic equations, high-order equations related to the Laplace operator and some elliptic systems [98]. We now shortly present two different approaches for constructing and justifying the WOS algorithm: the first, conventional, coming from approach (a), and the second based on a converse mean value relation.

Let us start with a simple case, the Dirichlet problem for the Laplace equation in a bounded domain $G \subset \mathbb{R}^3$:

$$\Delta u(x) = 0, \quad x \in G, \tag{1.1}$$

$$u(y) = \varphi(y), \quad y \in \Gamma = \partial G. \tag{1.2}$$

We seek a regular solution to (1.1) and (1.2), i.e., $u \in C^2(G) \bigcap C(G \bigcup \Gamma)$.

Let $d^* := \sup_{x\in G} d(x)$, where $d(x)$ is the largest radius of a sphere $S(x, d(x))$ centred at x and contained in $\overline{G}$.

Let $w_x(t)$ be the Wiener process starting at the point $x \in G$, and let τ_Γ be the first passage time (the time of the first intersection of the process $w_x(t)$ with the boundary Γ). We suppose that the boundary Γ is regular so that (1.1) and (1.2) have the unique solution. Then [19],

$$u(x) = \mathbb{E}_x \varphi(w_x(\tau_\Gamma)). \tag{1.3}$$

Note that in (1.3), only the random points on the boundary are involved. We thus can formulate the following problem: how to find these points without explicit simulation of the Wiener process inside the domain G?

This problem was solved in the study of Müller [76] using the following considerations. In the sphere $S(x, d(x))$, representation (1.3) gives

$$u(x) = \mathbb{E}_x u(w_x(\tau_{S(x,d(x))})). \tag{1.4}$$

The same representation is valid for all points $y \in S(x, d(x))$, so we can use the strong Markov property and write the conditional expectation:

$$u(x) = \mathbb{E}_x\{\mathbb{E}_y u(w_y(\tau_{S(y,d(y))}))|x = w(0); y = w(\tau_{S(x,d(x))})\}. \tag{1.5}$$

We can iterate this representation many times and remark that only random points lying on the spheres $S(x, d(x))$, $S(y, d(y))$, ..., are involved. It is well known that the points $w_x(\tau_{S(x,d(x))})$ are uniformly distributed over the sphere $S(x, d(x))$.

Thus, we came to the definition of the WOS process starting at x: it is defined as the homogeneous Markov chain $\mathrm{WS} = \mathrm{WS}\{x_0, x_1, \ldots, x_k, \ldots\}$ such that $x_0 = x$ and

$$x_k = x_{k-1} + d(x_{k-1})\omega_k, \quad k = 1, 2, \ldots,$$

where $\{\omega_k\}$ is a sequence of independent unit isotropic vectors. It is known [76] that $x_k \to y \in \Gamma$ as $k \to \infty$; however, the number of steps in the Markov chain WS is infinite with probability 1. Therefore, an ε-spherical process is introduced as follows. Let $N_\varepsilon = \inf\{n : d(x_n) \le \varepsilon\}$, then the ε-spherical process $\mathrm{WS}_\varepsilon = \{x_0, x_1, \ldots, x_{N_\varepsilon}\}$ is obtained from WS by stopping after N_ε steps.

Let $\bar{x}_{N_\varepsilon}$ be any point on $S(x_{N_\varepsilon}, d(x_{N_\varepsilon})) \cap \Gamma$, then we set

$$\hat{\xi}(x) = u(x_{N_\varepsilon}), \quad \xi(x) = \varphi(\bar{x}_{N_\varepsilon}). \tag{1.6}$$

The unbiasedness $u(x) = \mathbb{E}_x \hat{\xi}(x)$ follows from (1.4), (1.5). An important question is what happens when $\varepsilon \to 0$. We remark that

$$\mathbb{E}\{u(x_{n+1})|x_1, \ldots, x_n\} = \mathbb{E}\{u(x_{n+1})|x_n\} = u(x_n)$$

by the spherical mean value theorem. It means that $\{u(x_0),\dots,u(x_n),\dots\}$ is a martingale with respect to $x_1,\dots,x_n,\dots$. Since $\{u(x_0),\dots,u(x_{N_\varepsilon})\}$ is a stopped process constructed from $\{u(x_0),\dots,u(x_n),\dots\}$, it is also a martingale with respect to $x_1,\dots,x_n$ [75]. Especially,

$$\mathbb{E}\{\hat{\xi}(x)|x_1,\dots,x_k;N_\varepsilon \geq k\} = \mathbb{E}\{u(x_{N_\varepsilon})|x_k,N_\varepsilon \geq k) = u(x_k),$$

and

$$\mathbb{E}_x\hat{\xi}(x) = u(x).$$

Let

$$L = \sup_{x,y\in G} ||x-y||, \quad M = \sup_{x\in\Gamma} \varphi(x)$$

and assume that $u(x)$ satisfies the condition

$$|u(x)-u(y)| \leq c||x-y||, \quad x,y \in \overline{G}.$$

It is not difficult to show [75] that

$$\mathbb{E}_x\{(\hat{\xi}(x)-u(x))^2\} \leq c^2 L,$$

and for the estimator $\xi(x)$, the following estimation is true:

$$\mathbb{E}_x(\xi-u)^2 \leq c^2\varepsilon^2 + 4Mc\varepsilon + c^2L^2,$$

and

$$\mathbb{E}_x\left\{\frac{1}{n}\sum_{k=1}^{n}\xi_k(x)-u(x)\right\}^2 \leq c^2\varepsilon^2 + \frac{1}{n}(4Mc\varepsilon + c^2L^2),$$

where ξ_n are independent ξ-estimators of $u(x)$.

We now turn to the second approach.

It is well known that every solution to (1.1) satisfies the mean value relation:

$$u(x) = N_r u(x) := \frac{1}{\omega_m}\int_{S(x,r)} u(x+rs)\,d\Omega(s) \tag{1.7}$$

for each $x \in G$ and for all spheres $S(x,r)$ contained in $\overline{G}$; ω_m is the surface area of the unit sphere $S(x,1)$ in $\mathbb{R}^m$. Besides, mean value relation (1.7) characterizes the solutions to (1.1). We need a stronger result, which presents an equivalent formulation of the problem (1.1), (1.2).

Proposition 1.1 (Integral Formulation). *We suppose that problem* (1.1), (1.2) *has a unique solution for any continuous function* φ. *Suppose that there exists a function* $v \in C(G \bigcup \Gamma)$, $u|_\Gamma = \varphi$, $v(x) = u(x)$ *in* Γ_ε $(\varepsilon \geq 0)$ *such that the mean value relation holds at every point* $x \in G \setminus \Gamma_\varepsilon$ *for the sphere* $S(x, d(x))$. *Then* $v(x)$ *is the unique solution to the problem* (1.1), (1.2).

Proof. The proof uses the maximum property [12]. Since u also satisfies the mean value relation for every sphere contained in G, we conclude that the function $u - v$ satisfies the mean value relation at every point $x \in G \setminus \Gamma_\varepsilon$. Let F be a closed set of points $x \in G \setminus \Gamma_\varepsilon$ where $u - v$ attains its maximum M. Let x_0 be a point of F, which has the minimal distance from the surface $Y = \{y : d(y) = \varepsilon\}$. If x_0 were an interior point of $x \in G \setminus \Gamma_\varepsilon$, we could find a sphere $S(x_0, r_0) \subset G$ for which the mean value relation holds, thus $u - v = M$ inside $S(x_0, r_0)$. Therefore, x_0 should belong to Y. We repeat the same argument for the minimum of $u - v$. Since $(u - v)|_Y = 0$, we have $v \equiv u$ in G. □

Using this proposition, we rewrite an integral equation equivalent to the problem (1.1), (1.2).

Let $\delta_x(y)$ be a generalized density describing the uniform distribution on a sphere $S(x, d(x))$, and we define the kernel function as follows:

$$k_\varepsilon(x, y) = \begin{cases} \delta_x(y) & \text{if } x \in G \setminus \Gamma_\varepsilon, \\ 0 & \text{if } x \in \Gamma_\varepsilon. \end{cases} \tag{1.8}$$

We define also a function $f_\varepsilon(x)$ in G:

$$f_\varepsilon(x) = \begin{cases} 0 & \text{if } x \in G \setminus \Gamma_\varepsilon, \\ u(x) & \text{if } x \in \Gamma_\varepsilon. \end{cases}$$

By the proposition, we can write an equivalent integral equation:

$$u(x) = K_\varepsilon u(x) + f_\varepsilon(x), \tag{1.9}$$

where the integral operator K_ε is defined by

$$K_\varepsilon \psi(x) = \int_G k_\varepsilon(x, y) \psi(y) dy \tag{1.10}$$

for each $\psi(x) \in C(G)$.

Proposition 1.2. *For any* $\varepsilon > 0$, *integral equation* (1.9) *has the unique solution given by*

$$u(x) = f_\varepsilon(x) + \sum_{i=1}^{\infty} K_\varepsilon^i f_\varepsilon(x). \tag{1.11}$$

Besides, this solution coincides with the solution to (1.1), (1.2).

Proof. Let ε be fixed. To show the convergence of the series

$$f_\varepsilon(x) + \sum_{i=1}^{N} K_\varepsilon^i f_\varepsilon(x), \tag{1.12}$$

it is sufficient to prove that $\|K_\varepsilon^2\|_{L_\infty} < 1$ (this fact also implies the uniqueness of the solution to (1.11)). Let $\nu(\varepsilon) = \varepsilon^2/4(d^*)^2$. For $x \in G \setminus \Gamma_\varepsilon$, we have

$$\begin{aligned}\int_G k_\varepsilon(x,y) \int_G k_\varepsilon(y,y')dy'\,dy &= \int_{G\setminus\Gamma_\varepsilon} \delta_x(y)\left(\int_G \delta_y(y')dy'\right)dy \\ &= \int_{G\setminus\Gamma_\varepsilon} \delta_x(y)dy \le 1 - \nu(\varepsilon) < 1.\end{aligned}$$

Let $v(x) := f_\varepsilon(x) + \sum_{i=1}^{\infty} K_\varepsilon^i f_\varepsilon(x)$. It is clear that v satisfies

$$v(x) = K_\varepsilon v(x) + f_\varepsilon(x).$$

□

Now we remark that, to use (1.9) for numerical purposes, it is necessary to know the solution to the boundary problem in Γ_ε. However, we obtain an approximation if we take in Γ_ε

$$u(x) = \bar{f}_\varepsilon(x) \approx \varphi(\bar{x}), \quad x \in \Gamma_\varepsilon, \tag{1.13}$$

where $\bar{x}$ is the point of Γ nearest to x, since $u \in C(G \bigcup \Gamma)$. Instead of (1.13), we could use any continuous extension of φ to Γ_ε (the ideal case is the harmonic extension).

Thus, let us consider the equation

$$u_\varepsilon(x) = K_\varepsilon u_\varepsilon(x) + \bar{f}_\varepsilon(x), \quad x \in G \setminus \Gamma_\varepsilon, \tag{1.14}$$

where $\bar{f}_\varepsilon$ is an approximation for $u(x)$ in Γ_ε, e.g. given by (1.13). Note that the solution to (1.14) is not harmonic, but it is clear that $|u(x) - u_\varepsilon(x)| = O(\omega(\varepsilon))$ as $\varepsilon \to 0$, where $\omega(\varepsilon)$ is the continuity modulus of the function $u(x)$ in Γ_ε. For example, for Lipschitz continuous functions, we get

$$|u(x) - u_\varepsilon(x)| \le C\varepsilon.$$

Having (1.14) and the convergence of the Neumann series, it is possible to use the standard Monte Carlo estimators for the integral equations of the second kind [23]. If we choose the transitional density in each sphere in accordance with the kernel (i.e., uniformly on the surface of a sphere) and introduce absorption in Γ_ε with probability 1 (no absorption in $G \setminus \Gamma_\varepsilon$), we exactly obtain the unbiased estimators $\hat{\xi}(x)$ and $\xi(x)$. Estimations for the variance of the estimators of the solutions to the integral equations can be obtained from the analysis of the kernel k_ε^2/p, where p is a transitional density.

This kernel in our case coincides with the kernel of the original integral equation, which leads to the convergence of the Neumann series representing the variance.

Thus, we see that in the WOS algorithms, we have no necessity to simulate long trajectories of the Wiener process. Nevertheless, we have to construct a series of random points distributed inside the domain.

However, we would like to find probabilistic representations in the form of an expectation taken over Markov chains defined on the boundary of the domain. We will see that this is really possible to do using the boundary integral equations of the potential theory and special Monte Carlo estimators for solving integral equations.

We now shortly present the idea of the walk-on-boundary algorithm for solving (1.1), (1.2). Suppose for simplicity that the domain is convex and the solution can be represented in the form of a double-layer potential [120],

$$u(x) = \int_{\Gamma} \frac{\cos \varphi_{yx}}{2\pi |x-y|^2} \mu(y) d\sigma(y), \tag{1.15}$$

where φ_{yx} is the angle between the vector $x - y$ and n_y, the **interior** normal vector at the point $y \in \Gamma$, and $\mu(y)$ is a continuous function satisfying the boundary integral equation:

$$\mu(y) = -\int_{\Gamma} p(y, y')\mu(y')d\sigma(y') + \varphi(y), \tag{1.16}$$

where

$$p(y, y') = \frac{\cos \varphi_{y'y}}{2\pi |y-y'|^2}.$$

It is clear that

$$d\Omega_x = \frac{d\sigma(y) \cos \varphi_{yx}}{|x-y|^2}$$

is a solid angle of view of the surface element $d\sigma(y)$ from the point $x \in G$. Thus, the isotropic distribution of y in the angular measure Ω_x at the point x corresponds to the distribution of y on Γ with the density

$$p_0(x, y) = \frac{\cos \varphi_{yx}}{4\pi |x-y|^2}. \tag{1.17}$$

Since G is convex, $p_0 \geq 0$ and for arbitrary $x \in G$

$$\int_{\Gamma} p_0(x, y) d\sigma(y) = 1.$$

Thus, we can rewrite (1.15) in the form of the expectation

$$u(x) = 2\,\mathbb{E}_x\,\mu(Y_0), \tag{1.18}$$

where Y_0 is a random point on Γ, which is obtained as the intersection of the ray $A_x = \{z\ : z = x + t\omega\ ;\ t \geq 0\}$ (ω is a unit isotropic vector), having random isotropic direction ω, with the boundary Γ. Note that if we could find $\mu(y)$ on Γ, we could use the representation (1.18) for numerical purposes. In Chapter 3, we shall see that $\mu(y)$ can be found from (1.16) by special iterations, which leads to the following algorithm.

Let us define the following Markov chain on the boundary: $WB_x\{Y_0, Y_1, \dots, Y_m\}$, where the first point Y_0 is obtained as described above, and

$$Y_{n+1} = Y_n + |Y_{n+1} - Y_n|\omega_n,$$

where $\{\omega_k\}_{k=0}^{m-1}$ is a sequence of independent unit isotropic vectors in $\mathbb{R}^3$. Based on the process $WB_x\{Y_0, Y_1, \dots, Y_m\}$, we can construct the following random estimator ($m \geq 2$):

$$\xi_m(x) = 2[\varphi(Y_0) - \varphi(Y_1) + \varphi(Y_2) + \dots] + (-1)^{m+1}\varphi(Y_m). \tag{1.19}$$

In Chapter 3, we will show that

$$u(x) \approx \mathbb{E}\xi_m,$$

and the larger m, the higher is the accuracy of this representation. More exactly, to achieve the accuracy ε, the cost is $O\left(\frac{|\ln\varepsilon|^2}{\varepsilon^2}\right)$, which shows that this algorithm has a high efficiency.

Remarkably, the same walk-on-boundary process can be used to solve interior and exterior Dirichlet, Neumann and the third BVPs. Note also that the method is grid free but gives the solution simultaneously in arbitrary points prescribed. When comparing with the classical probabilistic representations, there is no ε-error coming from the approximation in Γ_ε. Note also that the parallelization of the walk-on-boundary algorithms is very simple, since the length of all the trajectories is one and the same.

In this book, we construct and justify the walk-on-boundary algorithms for three classes of PDEs: (1) classical stationary potential problems (Chapter 3), (2) heat equation (Chapter 4) and (3) spatial elasticity problems (Chapter 5). The basic Monte Carlo methods using Markov chains for solving integral equations are presented in Chapter 2. In Chapter 6, we discuss different aspects related to the walk-on-boundary algorithms (the Robin problem and the ergodicity theorem, evaluation of derivatives, etc.).

2 Random walk algorithms for solving integral equations

Conventional Monte Carlo methods for solving integral equations of the second kind are based on the Neumann series' representation of the solution and, consequently, they are applicable only if simple iterations are convergent. This, in turn, happens when the spectral radius of the integral operator is less than 1 (more exact formulations are given in the next section). However, boundary integral equations of the potential theory cannot be treated in the framework of this conventional scheme, since the spectral radii of these equations are no less than 1. Therefore, we shall introduce different, in general biased estimators for solving integral equations whose Neumann series may be divergent. This technique can be applied to general integral equations with completely continuous operators.

2.1 Conventional Monte Carlo scheme

Consider an integral equation of the second kind,

$$\varphi(x) = \int_{\Gamma} k(x,x')\varphi(x')dx' + f(x), \tag{2.1}$$

or in the operator form,

$$\varphi = K\varphi + f.$$

Here, Γ is a bounded domain in $\mathbb{R}^m$.

We suppose that the integral operator K is defined in a Banach space $X(\Gamma)$ of functions integrable in the domain Γ. For example, if

- $\{\int_{\Gamma} |k(x,x')|^r dx'\}^{1/r} \le C_1$ for almost all $x \in \Gamma$, $r > 0$,
- $\{\int_{\Gamma} |k(x,x')|^{\sigma} dx\}^{1/\sigma} \le C_2$ for almost all $x' \in \Gamma'$, $\sigma > 0$,
- $p > \sigma > p - r(p-1)$,

then K is a completely continuous operator from $L^p(\Gamma)$ to $L^p(\Gamma)$ with the norm [47]

$$||K|| \le C_1^{1-\sigma/p}\, C_2^{\sigma/p}.$$

In particular, this is true for the kernels of the potential theory:

$$k(x,x') = \frac{b(x,x')}{|x-x'|^n},$$

where $b(x, x')$ is a bounded continuous function, $n < m$. If, in addition, $p > m/(m-n)$, then $K : L^p(\Gamma) \to C(\Gamma)$.

We now define a Markov chain in the phase space Γ as a sequence of random points $Y = \{Y_0, Y_1, \dots, Y_n, \dots\}$, where the starting point Y_0 is distributed in Γ with an initial density $p_0(x)$, and the next points are determined by the transitional density

$$p(Y_{i-1} \to Y_i) = r(Y_{i-1}, Y_i)(1 - g(Y_{i-1})).$$

Here $r(Y_{i-1}, Y_i)$ is a conditional distribution density of Y_i, Y_{i-1} fixed, and $g(Y_{i-1})$ is the probability that the chain terminates in the state Y_{i-1} ($i \geq 1$). Thus, depending on the value of g, the chain may have infinite or finite number, N, of states.

Based on the Markov chain Y, we define the following random weights: the initial one

$$Q_0 = \frac{f(Y_0)}{p_0(Y_0)},$$

and

$$Q_i = Q_{i-1} \frac{k(Y_i, Y_{i-1})}{p(Y_{i-1}, Y_i)}, \quad i \geq 1.$$

It is supposed that p_0 and p are consistent with f and k, respectively, i.e.,

$$p_0(x) \neq 0 \text{ for } \{x : f(x) \neq 0\},$$

and

$$p(x, y) \neq 0 \text{ for } \{y : k(y, x) \neq 0\}.$$

These suppositions are necessary for the weights Q_i to be finite.

Our goal is to evaluate a linear integral functional

$$I_h = (\varphi, h) = \int_\Gamma \varphi(x) h(x) dx \tag{2.2}$$

for a function $h \in X^*(\Gamma)$. For example, if $h = \delta(x - x_0)$, then $I_\delta = \varphi(x_0)$ is the solution to (2.1) at the point x_0. The following statement is true [23].

Proposition 2.1. *Random variables*

$$\xi_i = Q_i h(Y_i) \tag{2.3}$$

are unbiased estimators for $(K^i f, h)$.

If the spectral radius of the integral operator

$$K_1 \varphi = \int_\Gamma |k(x, y)| \varphi(y) dy$$

is less than 1, then

$$\xi = \sum_{i=0}^{N} Q_i h(Y_i) \tag{2.4}$$

is an unbiased estimator for I_h; i.e.,

$$\mathbb{E}\xi_i = (K^i f, h), \tag{2.5}$$

$$\mathbb{E}\xi = I_h. \tag{2.6}$$

Proof. Define δ_n in the following way. We set $\delta_n = 0$ if the chain terminates at x_{n-1}, and $\delta_n = 1$ if the transition $x_{n-1} \to x_n$ happens. Next, let

$$\Delta_n = \prod_{k=0}^{n} \delta_k.$$

Thus, $\Delta_n = 1$ until the first termination of the chain.

Hence

$$\xi = \sum_{n=0}^{\infty} \Delta_n Q_n h(x_n).$$

Suppose for a while that functions $k(x', x), f$ and h are non-negative. Then, we can take expectation in the series termwise:

$$\mathbb{E}\xi = \sum_{n=0}^{\infty} \mathbb{E}[\Delta_n Q_n h(x_n)].$$

The conditional expectation formula gives

$$\begin{aligned}
\mathbb{E}[\Delta_n Q_n h(x_n)] &= \mathbb{E}_{(Y_0,\dots,Y_n)}\, \mathbb{E}\left[\Delta_n Q_n h(x_n) | Y_0, \dots, Y_n\right] \\
&= \mathbb{E}_{(Y_0,\dots,Y_n)}[Q_n h(Y_n)\, \mathbb{E}\left(\Delta_n | Y_0, \dots, Y_n\right)] \\
&= \mathbb{E}_{(Y_0,\dots,Y_n)}[Q_n h(Y_n) \prod_{k=0}^{n-1} (1 - g(x_n))] \\
&= \int dx_0 \int_{\Gamma} dx_n h(x_n) p_0(x_0) \left[\prod_{k=0}^{n-1} r(x_k, x_{k+1})\right] \left[\prod_{k=0}^{n-1} \frac{k(x_k, x_{k+1})}{p(x_k, x_{k+1})}\right] \\
&\quad \times \frac{f(x_0)}{p_0(x_0)} \left[\prod_{k=0}^{n-1} (1 - g(x_k))\right] \\
&= \int_{\Gamma} dx_0 \dots \int_{\Gamma} dx_n f(x_0) h(x_n) \left[\prod_{k=0}^{n-1} k(x_k, x_{k+1})\right] \\
&= (K^n f, h).
\end{aligned}$$

We have used here the obvious relation

$$\mathbb{E}[\Delta_n | Y_0, \ldots, Y_n] = \mathbb{P}(\Delta_n = 1 | Y_0, \ldots, Y_n)$$
$$= \mathbb{P}(\delta_0 = \delta_1 = \ldots = \delta_n = 1 | Y_0, \ldots, Y_n) = \prod_{k=0}^{n-1} [1 - g(x_k)].$$

We turn now to the general case of functions k, f and h. Denote by $Q_n^{(1)}$ the random weights constructed for

$$k_1(x', x) = |k(x', x)|, \quad f_1(x) = |f(x)|, \quad h_1(x) = |h(x)|.$$

Then

$$|\eta_m| = \left| \sum_{n=0}^{m} \Delta_n Q_n h(Y_n) \right| \le \sum_{n=0}^{m} |\Delta_n Q_n h(Y_n)|$$
$$= \sum_{n=0}^{m} \Delta_n Q_n^{(1)} h_1(Y_n) = \eta_m^{(1)},$$

and

$$\eta_m^{(1)} \to \xi_1 = \sum_{n=0}^{\infty} \Delta_n Q_n^{(1)} h_1(x_n).$$

By the assumptions made, $\mathbb{E}\xi_1 = (\varphi_1, h_1) < \infty$ (here $\varphi_1 = K_1\varphi_1 + f_1$).
By the Lebesgue theorem,

$$\lim_{m\to\infty} \mathbb{E}\eta_m = \mathbb{E}(\lim_{m\to\infty} \eta_m) = \mathbb{E}\xi.$$

But

$$\mathbb{E}\eta_m = \mathbb{E}\sum_{k=0}^{m} \Delta_k Q_k h(Y_k) = \sum_{k=0}^{m} \mathbb{E}(\Delta_k Q_k h(Y_k)) = \sum_{k=0}^{m} (K^k f, h).$$

Thus,

$$\mathbb{E}\xi = \lim_{m\to\infty} \mathbb{E}\eta_m = \sum_{n=0}^{\infty} (K^n f, h) = I_h,$$

since

$$||K^{n_0}|| \le ||K_1^{n_0}|| < 1$$

for some positive n_0. □

Note that it is possible to construct adjoint random estimators also. Let

$$Q_0^* = \frac{h(Y_0)}{p_0(Y_0)},$$
$$Q_i^* = Q_{i-1}^* \frac{k(Y_{i-1}, Y_i)}{p(Y_{i-1}, Y_i)}, \quad i \geq 1,$$

and

$$\xi_i^* = Q_i^* f(Y_i),$$
$$\xi^* = \sum_{i=0}^{N} Q_i^* f(Y_i). \tag{2.7}$$

Then, using the relations

$$(K^i f, h) = (f, K^{*i} h),$$

and

$$(\varphi, h) = (f, \varphi^*),$$

where $\varphi^* = K^* \varphi^* + h$,

$$K^* \varphi^*(x) = \int_\Gamma k(y, x) \varphi^*(y) dy,$$

we can prove that

$$\mathbb{E}\xi_i^* = \mathbb{E}\xi_i = (K^i f, h), \tag{2.8}$$

and

$$\mathbb{E}\xi^* = \mathbb{E}\xi = I_h. \tag{2.9}$$

Taking $f(x') = \delta(x - x')$, $p_0(x') = \delta(x - x')$, $Q_0 = 1$, we get

$$\varphi^* = h(x) + \mathbb{E} \sum_{n=1}^{N} Q_n h(Y_n), \tag{2.10}$$

which can be proved by using the representation

$$\varphi^* = \sum_{n=0}^{\infty} K^{*^n} h.$$

Or, considering the initial equation as an adjoint to $\varphi^* = K^* \varphi^* + h$, we get

$$\varphi = f + \mathbb{E} \sum_{n=0}^{N} Q_n^* f. \tag{2.11}$$

Variance of ξ can be represented as

$$\mathbb{D}\xi = (\chi, [2\varphi^* - h]) - I_h^2, \tag{2.12}$$

where χ is the Neumann series for the equation

$$\chi(x) = \int_{\Gamma} \frac{k^2(x, x')}{p(x', x)} \chi(x') dx' + \frac{f^2(x)}{p_0(x)}. \tag{2.13}$$

Thus, if $||K_1^{n_0}|| < 1$ for some integer number n_0 and the variance of the direct estimator for the integral equation with the kernel $|k|$ is finite, then the variance $\mathbb{D}\xi$ is finite too.

Remark 2.2. Let us consider a system of integral equations

$$\varphi_i(x) = \sum_{j=1}^{s} k_{ij}(x, y)\varphi_j(y)dy + f_i(x), \; i = 1, \dots, s$$

or in the matrix-operator form

$$\Phi(x) = \int_{\Gamma} K[x, y]\Phi(y)dy + F(x).$$

Suppose it is desired to find the linear functional

$$J(x) = \int_{\Gamma} P[x, y]\Phi(y)dy \equiv (P, \Phi).$$

Here, $K[x, y]$ is the matrix of kernels $\{K_{ij}\}_{i,j=1}^{s}$, $\Phi = (\varphi_1, \dots, \varphi_s)^T$, and $P[x, y]$ is a given matrix.

It is not difficult to extend the estimators ξ_i, ξ_i^* for the iterations $(P, \mathcal{K}^i[\,]F)$, where $\mathcal{K}[\,]$ is the matrix-integral operator generated by the matrix $K[x, y]$:

$$\mathcal{K}[\,]F = \int_{\Gamma} K[x, y]F(y)dy.$$

Indeed, let us choose a homogeneous Markov chain $\{Y_0, Y_1, \dots\}$ defined by the initial density p_0 and the transitional density $p(x, y)$, such that

$$p_0(y) \neq 0 \text{ for } y : P[x, y]\Phi(y) \not\equiv 0,$$

and

$$p(x,y) \neq 0 \text{ for } y : K[x,y]\Phi(y) \not\equiv 0.$$

Then the random variables

$$\xi_i = P[x, Y_i]Q_i, \quad \xi_i^\star = Q_i^\star F(Y_i)$$

are unbiased estimators:

$$\mathbb{E}\xi_i = \mathbb{E}\xi_i^\star = (P, K^i[\,]F).$$

Here, Q_i are vector weights

$$Q_i = \frac{K[Y_i, Y_{i-1}]Q_{i-1}}{p(Y_{i-1}, Y_i)}, \quad Q_0 = \frac{F(Y_0)}{p_0(Y_0)},$$

and $Q_i^\star$ are matrix weights:

$$Q_i^\star = Q_{i-1}^\star \frac{K[Y_{i-1}, Y_i]}{p(Y_{i-1}, Y_i)}, \quad Q_0^\star = \frac{P[x, Y_0]}{p_0(Y_0)}.$$

2.2 Biased estimators

Let us consider an equation of type (2.1), introducing a complex variable λ:

$$\varphi(x) = \lambda K\varphi + f. \tag{2.14}$$

Let $\lambda \neq 0$ be a non-singular value. This means there exists a unique solution to (2.14). Then, the operator R_λ defined from

$$I + \lambda R_\lambda = (I - \lambda K)^{-1}$$

is called a resolvent operator. Here, I denotes the identity operator.

Let $\chi_0(K)$ be the set of characteristic numbers of the operator K, i.e., the set of such values of λ that the equation $\varphi = \lambda K\varphi$ has non-zero solutions. We suppose that K is a completely continuous operator. Then, the set $\chi_0(K) = \{\lambda_1, \lambda_2, \dots\}$ is countable and we assume that $|\lambda_1| \le |\lambda_2| \le \dots$.

Without loss of generality, we suppose that it is necessary to find the solution to (2.14) at $\lambda = \lambda_\star = 1 \notin \chi_0(K)$.

If $|\lambda_1| \le 1$, then the Neumann series diverges at $\lambda_\star = 1$ and the conventional scheme described in Section 2.1 fails. However, the function $R_\lambda f$ is analytic in the circle $\{\lambda : |\lambda| < |\lambda_1|\}$ and can be analytically continued to any regular λ, and, in particular, to the point $\lambda_\star = 1$. To carry out this continuation, we use the conformal mapping of the parameter λ [46].

Let us take a simple connected subdomain D in the complex plane such that $D \cap \chi_0(K) = \emptyset$, and $\lambda_* \in D$, $0 \in D$. We choose a conformal mapping $\lambda = \psi(\eta)$, $\psi(0) = 0$, which maps the unit disk

$$\Delta = \{\eta : |\eta| < 1\} \tag{2.15}$$

onto D. Then

$$\eta_* = \psi^{-1}(\lambda_*) \in \Delta,$$

and $\psi^{-1}(\lambda_k) \notin \Delta$.

We use now the expansion

$$\lambda^i = \sum_{k=1}^{\infty} d_i^{(k)} \eta^k,$$

where

$$d_i^{(k)} = \frac{1}{k!} \left\{ \frac{\partial^k}{\partial \eta^k} [\psi(\eta)]^i \right\}_{\eta=0}.$$

Substitution of $\lambda = \psi(\eta)$ into $\lambda R_\lambda f$ gives $F(\eta) = \psi(\eta) R_{\psi(\eta)} f$, and after the expansion into series

$$F(\eta) = \sum_{k=1}^{\infty} b_k \eta^k,$$

where

$$b_k = \sum_{i=1}^{k} d_i^{(k)} K^i f.$$

Thus, the series

$$\varphi(x) = f(x) + \sum_{k=1}^{\infty} \left\{ \sum_{i=1}^{k} d_i^{(k)} K^i f(x) \right\} \eta_*^k \tag{2.16a}$$

is absolutely and uniformly convergent.

Let

$$\eta_0 = \min_k |\psi^{-1}(\lambda_k)| \geq 1.$$

We cut off the series, taking first n terms:

$$\varphi(x) = f(x) + \sum_{i=1}^{n} l_i^{(n)} K^i f(x) + \varepsilon(n; x), \tag{2.16b}$$

where

$$l_i^{(n)} = \sum_{k=1}^{n} d_i^{(k)} \eta_*^k, \tag{2.17}$$

$$|\varepsilon(n;x)| \le \text{const} \cdot \left|\frac{|\eta_*|}{\eta_0}\right|^n .$$

We assume that the region D and the mapping $\psi(\eta)$ can be chosen so that all the coefficients $l_i^{(n)}$ and the value η_* can be calculated with the desired accuracy.

To evaluate I_h, we use the unbiased estimators ξ_i defined in (2.3) or ξ_i^* defined in (2.7). Then, we can introduce the random estimators

$$\xi(n) = \sum_{i=0}^{n} l_i^{(n)} Q_i h(Y_i),$$
$$\xi^*(n) = \sum_{i=0}^{n} l_i^{(n)} Q_i^* f(Y_i). \tag{2.18}$$

By the construction, these estimators are δ-biased:

$$I_h = \mathbb{E}\xi(n) + \delta_1,$$
$$I_h = \mathbb{E}\xi^*(n) + \delta_2, \tag{2.19}$$

where $\delta_1 = (\varepsilon(n;\cdot), h)$, $\delta_2 = (f, \varepsilon^*(n;\cdot))$, $l_0^{(n)} = 1$, $g(Y_i) = 0$ for $i < n$ and $g(Y_n) = 1$.

Further, for $\xi(n)$, we use the name 'direct δ-estimator', and for $\xi^*(n)$, we use the name 'adjoint δ-estimator' for I_h.

2.3 Linear-fractional transformations and their relations to iterative processes

Assume that a solution to the integral equation (2.14) with a completely continuous operator K can be represented in form (2.16a). We rewrite it as

$$\varphi(x) = f(x) + \sum_{k=1}^{\infty} \varphi_k(x) \eta_*^k, \tag{2.20}$$

where

$$\varphi_k(x) = \sum_{i=1}^{k} d_i^{(k)} K^i f(x). \tag{2.21}$$

Introduce the notation

$$\varphi^{(m)}(x) = f(x) + \sum_{k=1}^{m} \varphi_k(x)\eta_*^k. \tag{2.22}$$

Our goal is to study the iterative process for $\varphi^{(m)}$ generated by the linear fractional transformation (α, β are complex parameters):

$$\lambda = \frac{\alpha\eta}{1-\beta\eta}, \tag{2.23}$$

which maps the unit disk Δ onto the half-plane:

$$D = \{\lambda : \Re\left[-\frac{\alpha}{\beta}|\bar{\lambda} - \bar{\lambda}_0|\right] < 0\} = \left\{\lambda : |\lambda| < |\beta|\left|\lambda + \frac{\alpha}{\beta}\right|\right\},$$

and

$$\lambda_0 = -\frac{\alpha|\beta|}{\beta(1+|\beta|)}.$$

In this case,

$$d_i^{(k)} = \alpha^i C_{k-1}^{i-1}\beta^{k-i}, \quad k \geq i,$$
$$\eta_* = (\alpha+\beta)^{-1}.$$

Note that (2.20) can be rewritten then as follows:

$$\varphi(x) = f(x) + \sum_{i=1}^{\infty} K_1^i f_1(x), \tag{2.24}$$

where

$$K_1 = \frac{\beta}{\alpha+\beta}I + \frac{\alpha}{\alpha+\beta}K, \qquad f_1 = \frac{\alpha}{\alpha+\beta}Kf. \tag{2.25}$$

Consequently, we get

$$\varphi_0(x) = \alpha Kf, \quad \varphi_k(x) = (\beta I + \alpha K)\varphi_{k-1}(x),$$
$$\varphi^{(m)}(x) = K_1\varphi^{(m-1)}(x) + \frac{\alpha}{\alpha+\beta}f(x),$$
$$\varphi^{(0)}(x) = f(x) + \frac{\alpha}{\alpha+\beta}Kf(x). \tag{2.26}$$

Thus, (2.24) shows that we can pass to a new integral equation with the integral operator K_1 in (2.25) and the right-hand side f_1. Assume that $\lambda_i(K) \notin D$. Then the series in (2.24) converges. The respective unbiased estimators for I_h are

$$\xi = \frac{f(x_0)h(x_0)}{p_0(x_0)} + \sum_{k=0}^{N} Q_k h(x_k),$$
$$\xi^* = \frac{f(x_0)h(x_0)}{p_0(x_0)} + \sum_{k=0}^{N} Q_k^* f_1^*(x_k), \tag{2.27}$$

where weights Q_k, Q_k^* are determined as described above with changing K by K_1 and f by f_1. The transitional density is taken in accordance with K_1. The upper number N in (2.27) can be taken fixed, or at random, if the Neumann series for $\overline{K}_1$ converges. Here $\overline{K}_1$ is the integral operator with the kernel $|k_1|$. The variance, however, may be unbounded.

Remark 2.3. Note that we could derive the iterative process by the following transformation of the original integral equation. Taking $a \neq 0$, we get

$$\varphi = [(1-a)I + aK]\varphi + af,$$

where $a = \alpha/(\alpha+\beta)$. The iterative process

$$\varphi^{(m)}(x) \overset{m\to\infty}{\to} \varphi(x)$$

is constructed as in (2.26), but with

$$\varphi^{(0)}(x) = \frac{\alpha}{\alpha+\beta} f(x).$$

Note also that the domain of convergence of the iterative process (2.26) is broader than D. Let, for example, $\beta = 1$, $\alpha > 0$. Then $D = \{\lambda : \Re\lambda > -\alpha/2\}$.

The Neumann series (2.24) converges if

$$|\psi^{-1}(\lambda_k)| > |\eta^*|, \quad k = 1, 2, \dots,$$

where $\{\lambda_k\}$ are the characteristic numbers of K. Let $\lambda = x + iy$, where $x = \Re\lambda$ and i is the imaginary unit. Then, we can rewrite this condition in the form

$$\left(x - \frac{1}{\alpha+2}\right)^2 + y^2 > \left(\frac{1+\alpha}{2+\alpha}\right)^2.$$

Consider now the transformation $\lambda = \alpha/(1-\beta\eta)$. It maps the disk Δ onto the half-plane:

$$D = \{\lambda : |\lambda - \alpha| < |\beta||\lambda|\},$$

and $\psi(0) \neq 0$,

$$d_i^{(k)} = \alpha^i C_{i+k-1}^{i-1} \beta^k, \quad \eta^* = (1-\alpha)/\beta.$$

This corresponds to iterations of the resolvent operator:

$$\varphi = (I - \alpha K)^{-1} f + \sum_{i=1}^{\infty} [(1-\alpha)(1-\alpha K)^{-1}]^i (I - \alpha K)^{-1} \alpha K f.$$

Parameter α is chosen so that the operator $(I - \alpha K)^{-1}$ exists. In this case, we can use the transformation ($\alpha \neq 1$):

$$\varphi = (1-\alpha)(I - \alpha K)^{-1} \varphi + (I - \alpha K)^{-1} \alpha f.$$

The iterative process generated by the transformation

$$\lambda = \frac{\alpha}{1 - \beta \eta}$$

has the form

$$\varphi^{(m)} = (1-\alpha)(I - \alpha K)^{-1} \varphi^{(m-1)} + (I - \alpha K)^{-1} \alpha f, \tag{2.28}$$

or

$$\varphi^{(m)} = \alpha K \varphi^{(m)} + (1-\alpha) \varphi^{(m-1)} + \alpha f.$$

The domain of convergence of this iterative process is

$$\left(x - \frac{1}{2-\alpha}\right)^2 + y^2 > \left(\frac{1-\alpha}{2-\alpha}\right)^2.$$

In this case, the Neumann series for the resolvent operator $(1-\alpha)(I - \alpha K)^{-1}$ converges.

Consider now the iterative process (2.26), (2.28) for fixed m, N. In this case, it is possible to change the order of summation; hence, the approximate solution is represented as an operator polynomial (of finite degree) applied to the right-hand side of the equation. Thus, for the series (2.24), we get

$$\varphi = \sum_{i=0}^{N} a_i(N) K^i f + \varepsilon(N), \tag{2.29}$$

where

$$a_0(N) = 1, \quad a_i(N) = a_i \sum_{k=i-1}^{N-1} C_k^{i-1} (1-a)^{k-i+1}, \tag{2.30}$$

$a = \frac{\alpha}{\alpha+\beta}$, N fixed.

If we take a finite number of terms in the expansion of the resolvent operator $[(1-\alpha)(I-\alpha K)^{-1}]^n$, then the iterative process (2.28) leads to the following expressions for the coefficients in (2.29):

$$a_i(N) = \alpha^{i+1} \sum_{k=1}^{N} C_k^i (1-\alpha)^{k-1}, \quad \text{where } N \le m-1, \text{ or } N \ge m;\ i > N-m+1,$$

and

$$a_i(N) = \alpha^i \sum_{k=1}^{i+m-1} C_k^i (1-\alpha)^{k-i}, \quad \text{where } N \ge m;\ i \le N-m+1. \tag{2.31}$$

Note, however, that for a particular case of K, it is possible to construct a non-biased estimator for the kernel of the operator R, defined by

$$(I-\alpha K)^{-1} = I + \alpha R.$$

This makes possible the construction of estimators for φ using double randomization and substituting into the weights not the exact value of the kernel $(I-\alpha K)^{-1}$ but its unbiased estimator. This estimator has a finite variance if the order of singularity of the kernel of K is less than $m/2$, where m is the dimension of the problem.

Consider a simple example, which will be used later on for construction of the walk-on-boundary algorithms. Assume that all the characteristic values are real, and $\lambda_k \in (-\infty, -a)$, $a > 0$. Without loss of generality, we assume that (2.1) is to be solved at $\lambda = \lambda_* = 1$. In this case, it is convenient to use for analytic continuation the following function that maps the disk Δ onto the domain D_a, the complex plane with a cut along the real axis from $-a$ to $-\infty$:

$$\lambda = \psi(\eta) = \frac{4a\eta}{(1-\eta)^2}. \tag{2.32}$$

The series in the expansion of $R_\psi(\eta)f$ converges absolutely and uniformly on Δ, and the coefficients can be exactly calculated:

$$d_k^{(n)} = (4a)^k C_{k+n-1}^{2k-1}.$$

To construct a biased random estimator, we choose such m that the remainder of the series is equal to a desired quantity ε. Then

$$R_{\lambda_*} f \cong \sum_{k=1}^{m} b_k \eta_*^k = \sum_{k=1}^{m} \eta_*^n \sum_{i=1}^{k} d_i^{(k)} c_i = \sum_{k=1}^{m} c_k l_k^{(m)}, \tag{2.33}$$

where

$$l_k^{(m)} = \sum_{n=k}^{m} d_k^{(n)} \eta_*^n. \tag{2.34}$$

The coefficients $l_k^{(m)}$ are calculated in advance, according to (2.34), and the integrals $c_k = K^{k+1}f$ are calculated by Monte Carlo methods. Let ξ_k be an unbiased Monte Carlo estimator for c_{k-1}. Hence, an ε-biased estimator for the solution to the integral equation has the form

$$\zeta_\varepsilon^{(m)}(x) = f(x) + \sum_{k=1}^{m} \xi_k l_k^{(m)}. \tag{2.35}$$

Theorem 2.1. *For the transformation (2.32), $l_k^{(m)} \le 1$ for all k and m. Let variances be bounded, $\mathbb{D}\xi_k \le \sigma^2$, and ε be the required order of the estimator $\zeta_\varepsilon^{(m)}$ error. Then, for $\varepsilon \to 0$, the computational cost of this estimator is*

$$T_\varepsilon = O(|\ln(\varepsilon)|^3/\varepsilon^2). \tag{2.36}$$

Proof. To prove the result, it is sufficient to show that $l_k^{(m)} \le 1$. Note that

$$\eta_* = \psi^{-1}(\lambda_*) = [(a+\lambda_*)^{1/2} - a^{1/2}]/[(a+\lambda_*)^{1/2} + a^{1/2}] > 0$$

and $d_k^{(n)} = (4a)^k c_{k+n-1}^{2k-1} > 0$, $n = k, k+1, \ldots$ Therefore, for a fixed arbitrary k, $l_k^{(m)}$ monotonically increases as m increases. In the limit $m \to \infty$, $l_k^{(\infty)} = 1$, since (2.34) gives 1^k, $k = 1, 2, \ldots$ From this, it follows that $|l_k^{(m)}| \le 1$ for arbitrary k and m. □

The following statement establishes an analogous result for a particular class of transformations.

Theorem 2.2. *Assume that the conformal mapping $\lambda = \psi(\eta)$ has only simple poles on the circle $|\eta| = 1$. Then, if $\mathbb{D}(k) \equiv \mathbb{D}\xi_k \le \sigma^2$, and if*

$$q = \frac{c|\eta_*|}{1 - |\eta_*|} < 1, \tag{2.37}$$

where c is the constant in the inequality $|a_n| \le c$, $a_n = d_1^n$, the cost of the estimator (2.35) is of order $(|\ln(\varepsilon)|^4/\varepsilon^2)$.

Proof. The estimate $|a_n| \le c$ can be obtained from the representation [83]:

$$\sum_{n=0}^{\infty} a_n z^n = \sum_{i=1}^{k} \frac{c_i}{1 - z_i z} + \sum_{n=0}^{\infty} b_n z^n,$$

where $|z_i| = 1$, $i = 1, \ldots, k$, $\lim_{n\to\infty} \sup |b_n|^{1/n} < 1$. From this, we get

$$|a_n| = \left|\sum_{i=1}^{k} c_i z_i^n + b_n\right| \le \sum_{i=1}^{k} |c_i| + B, \, |b_n| < B.$$

Therefore,

$$|\psi(\eta_*)| \le c\frac{|\eta_*|}{1 - |\eta_*|}.$$

Using the Cauchy inequality, we obtain

$$|d_k^{(n)}| \le c^k \frac{|\eta_*|^k}{(1 - |\eta_*|)^k |\eta_*|^n},$$

and consequently,

$$|l_k^{(m)}| \le \sum_{n=k}^{m} \frac{c^k |\eta_*|^k}{(1 - |\eta_*|)^k} = q^k(m - k + 1).$$

Thus,

$$\mathbb{D}\zeta_\varepsilon^{(m)}(x) \le m \sum_{n=k}^{M} (m - k + 1)^2 q^{2k}\, \mathbb{D}(k) < \frac{q^2 \sigma^2 m^2}{1 - q^2}.$$

The last inequality proves the theorem since $m = 0(\ln |\varepsilon|)$ due to the fact that (2.20) converges as a power series. □

Theorem 2.3. *Assume that a regular function $\lambda = \psi_B(\eta) = a_1\eta + a_2\eta^2 + \ldots$ maps the disk $|\eta| < 1$ onto a convex domain. Then, the condition (2.37) takes the form*

$$q = \frac{|\psi_B(\eta_*)|}{1 - |\eta_*|} < 1, \; a_1 = 1.$$

Let $\mathbb{D}(k) \le \sigma^2$. Then, asymptotically, the cost of the estimator is $O(|\ln(\varepsilon)|^4/\varepsilon^2)$.

Proof. The function $\psi_B(\eta)$ has the property $|a_n| \le 1$ [83]. To complete the proof, one only needs to use the arguments of the proof of Theorem 2.2. □

Remark 2.4. Changing assumptions about $\mathbb{D}(k)$ can lead to different cost estimations in Theorems 2.1–2.3.

Indeed, if $\mathbb{D}(k) = p^k \sigma^2$, for $p < 1$ (e.g. if $||K^2/r|| < 1$), then $\mathbb{D}\zeta_\varepsilon^{(m)} < c\sigma^2 m$ (if $l_k^{(m)} < c$), and $T_\varepsilon = O(|\ln(\varepsilon)|^2/\varepsilon^2)$. In this case, it is possible to assume $|pq^2| < 1$ instead of (2.37). For $p > 1$, $\mathbb{D}\zeta_\varepsilon^{(m)} < q^2\sigma^2 m^3/(1 - |pq^2|)$ if $|pq^2| < 1$, and $T_\varepsilon = O(|\ln(\varepsilon)|^4/\varepsilon^2)$.

Thus, to estimate the cost of (2.35), it is necessary to have more information about $l_k^{(m)}$ and $\mathbb{D}(k)$.

2.4 Asymptotically unbiased estimators based on singular approximations

We now consider a different approach to solving the integral equation

$$\varphi(x) = \int k(x, y)\varphi(y)dy + f(x). \tag{2.38}$$

The algorithms considered here are based on the well-known method [46], which exploits a singular approximation to the kernel $k(x, y)$. The solution to the equation to be solved is simply represented through linear functionals of auxiliary equations. This fact can be effectively exploited in the Monte Carlo method. The difference of the approach presented consists of the possibility to use a 'few-point approximation', in particular, a one-point approximation

$$k(x, y) = k(x, y_0)\gamma(y) + k_1(x, y). \tag{2.39}$$

It is required only that the Neumann series for the integral equation with the kernel $|k_1(x, y)|$ converges.

For the sake of simplicity, consider a system of linear algebraic equations, i.e., (2.38) of the form $x = Ax + b$, or in n coordinates

$$x_i = \sum_{j=1}^{n} a_{ij}x_j + b_i, \quad i = 1, \ldots, n. \tag{2.40}$$

Consider two auxiliary linear systems

$$\begin{aligned} z_i &= \sum_{j=1}^{n} d_{ij}z_j + b_i, \\ y_i &= \sum_{j=1}^{n} d_{ij}y_j + a_{ij_0}, \quad i = 1, \ldots, n \end{aligned} \tag{2.41}$$

where $d_{ij} = a_{ij} - a_{ij_0}\gamma_j$ are the entries of a matrix D, $\{a_{ij_0}, i = 1, \ldots, n\}$ is a fixed column of the matrix A, $\gamma^T = (\gamma_1, \ldots, \gamma_n)$ is an arbitrary vector.

Theorem 2.4. *Assume that systems (2.40), (2.41) are uniquely solvable and*

$$(\gamma, y) = \sum_{j=1}^{n} \gamma_i y_i \neq 1.$$

Then,

$$x_i = z_i + y_i \frac{\sum_{j=1}^{n} \gamma_j z_j}{1 - \sum_{j=1}^{n} \gamma_j y_j}, \quad i = 1, \ldots, n. \tag{2.42}$$

Proof. Let $T = (\gamma, x)$. We prove first that $x_i = z_i + y_i T$, $i = 1, \dots, n$. Indeed,

$$\begin{aligned} z_i + y_i T &= \sum_{j=1}^{n} d_{ij} z_j + b_i + T \sum_{j=1}^{n} d_{ij} y_j + T a_{ij_0} \\ &= b_i + a_{ij_0} T + \sum_{j=1}^{n} d_{ij}(z_j + y_j T). \end{aligned} \tag{2.43}$$

Substituting $a_{ij} = d_{ij} + a_{ij_0}\gamma_j$ into (2.40) yields

$$x_i = \sum_{j=1}^{n} d_{ij} x_j + b_i + a_{ij_0} T,$$

so we obtain the desired equality. We use it in the following transformations:

$$\begin{aligned} x_i &= z_i + y_i \sum_{j=1}^{n} \gamma_j x_j = z_i + y_i \sum_{j=1}^{n} \gamma_j (z_j + y_j T) \\ &= z_i + y_i \sum_{j=1}^{n} \gamma_j z_j + y_i T(\gamma, y) = z_i + y_i(\gamma, z) + (x_i - z_i)(\gamma, y) \\ &= z_i(1 - (\gamma, y)) + y_i(\gamma, z) + x_i(\gamma, y), \end{aligned}$$

and arrive at the desired equality

$$x_i = z_i + \frac{y_i(\gamma, z)}{1 - (\gamma, y)},$$

which proves the theorem. □

The equality (2.42) will be useful if the solution of (2.41) is somehow simpler than the construction of the solution of the original equation. In particular, the situation when $\varrho(A) > 1$ but $\varrho(D) < 1$ is interesting from the point of view of Monte Carlo methods, since in this case, it is possible to construct estimators for $z_i, y_i, (\gamma, z)$ and (γ, y) simultaneously, using a single Markov chain.

Consider now the case when the kernel has the form (2.39). Then the auxiliary equations are

$$\begin{aligned} \varphi_1(x) &= \int_G k_1(x, y)\varphi_1(y)dy + f(x), \\ \varphi_2(x) &= \int_G k_1(x, y)\varphi_2(y)dy + k(x, y_0), \end{aligned} \tag{2.44}$$

where y_0 is an arbitrary fixed point in G; $\gamma(y)$ is a function such that (γ, φ_1) and (γ, φ_2) exist.

Theorem 2.5. *Assume that equations (2.44) are uniquely solvable for arbitrary right-hand sides (from a class of functions), and suppose that*

$$(\gamma, \varphi_2) = \int_G \gamma(y)\varphi_2(y)dy \neq 1.$$

Then,

$$\varphi(x) = \varphi_1(x) + \varphi_2(x)\frac{\int_G \gamma(y)\varphi_1(y)dy}{1 - (\gamma, \varphi_2)}. \tag{2.45}$$

Proof. The proof is analogous to that of Theorem 2.4. Indeed, substituting (2.39) into (2.38) yields

$$\varphi(x) = f(x) + \int_G k_1(x, y)\varphi(y)dy + k(x, y_0)J, \tag{2.46}$$

where $J = (\gamma, \varphi) = \int_G \gamma(y)\varphi(y)dy$. We show now that $\varphi(x) = \varphi_1(x) + \varphi_2(x)J$. Indeed,

$$\begin{aligned}\varphi_1(x) + \varphi_2(x)J = &\int_G k_1(x, y)\varphi_1(y)dy + f(x) \\ &+ J\int_G k_1(x, y)\varphi_2(y)dy + k(x, y_0)J = f(x) + k(x, y_0)J \\ &+ \int_G k_1(x, y)[\varphi_1(y) + \varphi_2(y)J]dy.\end{aligned}$$

Taking into account (2.46), we obtain the desired equality. Like in Theorem 2.4, we use it in the following transformations:

$$\begin{aligned}\varphi(x) &= \varphi_1(x) + \varphi_2(x)J = \varphi_1(x) + \varphi_2(x)\int_G \gamma(y)\varphi(y)dy \\ &= \varphi_1(x) + \varphi_2(x)\int \gamma(y)[\varphi_1(y) + \varphi_2(y)J]dy \\ &= \varphi_1(x) + \varphi_2(x)(\gamma, \varphi_1) + \varphi_2(x)J(\gamma, \varphi_2) \\ &= \varphi_1(x) + \varphi_2(x)(\gamma, \varphi_1) + [\varphi(x) - \varphi_1(x)](\gamma, \varphi_2) \\ &= \varphi_1(x) + \varphi_2(x)(\gamma, \varphi_1) + \varphi(x)(\gamma, \varphi_2) - \varphi_1(x)(\gamma, \varphi_2) \\ &= \varphi_1(x)[1 - (\gamma, \varphi_2)] + \varphi(x)(\gamma, \varphi_2) + \varphi_2(x)(\gamma, \varphi_1)\end{aligned}$$

and the theorem is proved. □

Using (2.45), it is possible to construct an asymptotically unbiased estimator for $\varphi(x)$ provided that $\varrho(\overline{K}_1) < 1$, where $\overline{K}_1$ is the integral operator with the kernel $|k_1|$.

Let $\xi_1(x), \xi_2(x)$, and $\xi_{1_\gamma}(x), \xi_{1_\gamma}(x)$ be unbiased estimators for $\varphi_1(x)$, $\varphi_2(x)$, and $(\varphi_1(x), \gamma), (\varphi_2(x), \gamma)$, respectively. Note that due to (2.44), all these estimators could be constructed using a single Markov chain; moreover, $\xi_{1_\gamma}(x), \xi_{1_\gamma}(x)$ will differ only in the initial weights. Then,

$$\eta(x) = \xi_1(x) + \xi_2(x)\frac{\widetilde{\xi}^{(m)}_{2_\gamma}}{1 - \widetilde{\xi}^{(m)}_{2_\gamma}},$$

$$\widetilde{\xi}^{(m)}_{l_\gamma} = \frac{1}{m}\sum_{k=1}^{m} \widetilde{\xi}^{(k)}_{l_\gamma}, \quad l = 1, 2 \tag{2.47}$$

is an asymptotically unbiased estimator for $\varphi(x)$. Indeed, let us assume that $\widetilde{\xi}^{(i)}_{l_\gamma}$, $1 - \widetilde{\xi}^{(i)}_{2_\gamma}$, $i = 1, 2, \ldots, m$ are independent realizations of ξ_{1_γ}, $1 - \xi_{2_\gamma}$, respectively, constructed using a Markov chain. Then, by the well-known theorem for distribution of a function of random variables [88], the random quantity

$$\zeta = \sum_{i=1}^{m} \xi^{(i)}_{1_\gamma}\left[\sum_{i=1}^{m}(1 - \xi^{(i)}_{2_\gamma})\right]^{-1}$$

is asymptotically Gaussian with the mean $(\gamma, \varphi_1)/[1 - (\gamma, \varphi_2)]$ and the variance

$$\frac{1}{m}\left[\frac{(\gamma, \varphi_1)}{1 - (\gamma, \varphi_2)}\right]^2 \left\{\frac{\mathbb{D}\xi_{1_\gamma}}{(\gamma, \varphi_1)^2} - 2\frac{cov[\xi_{1_\gamma}(1 - \xi_{2_\gamma})]}{(\gamma, \varphi_1)[1 - (\gamma, \varphi_2)]} + \frac{\mathbb{D}(1 - \xi_{2_\gamma})}{[1 - (\gamma, \varphi_2)]^2}\right\}.$$

Analogous arguments can be used for the random variable
$\xi_2(x)\widetilde{\xi}^{(m)}_{1_\gamma}/[1 - \widetilde{\xi}^{(m)}_{2_\gamma}]$; hence, η is an asymptotically unbiased estimator for $\varphi(x)$.

The approach described above can be generalized to systems of integral equations. Systems of linear algebraic equations can also be treated.

Consider a system of integral equations

$$\varphi(x) = \int_G K[x, y]dy + f(x), \tag{2.48}$$

where $f(x)$ is a known vector; $K[x, y]$ is a matrix with elements $\{k(x, y)\}_{i,j=1}^{m}$, and $\varphi(x) = (\varphi^1(x), \ldots, \varphi^m(x))^T$ is the column vector of functions to be found. No special assumptions about the properties of the integral operator $K[\ ]$, the class of vector functions φ, f, and the domain G will be used. We assume only that the system (2.48) is uniquely solvable and generalize the representation (2.42).

First, for simplicity, consider the case $n = 1$, i.e.,

$$M[x, y] = K[x, y] - \delta(x)\gamma^T(y),$$

or in more detail

$$M_{ij}(x,y) = k_{ij}(x,y) - \delta^i(x)\gamma^i(y), \tag{2.49}$$

where $\delta(x)$ and $\gamma(y)$ are some arbitrary column vectors with components $\delta^1(x),\dots,\delta^m(x)$ and $\gamma^1(y),\dots,\gamma^m(y)$, respectively.

We introduce two auxiliary systems of integral equations

$$\begin{aligned}\psi_0(x) &= \int_G M[x,y]\psi_0(y)dy + f(x),\\ \psi_1(x) &= \int_G M[x,y]\psi_1(y)dy + \delta(x).\end{aligned} \tag{2.50}$$

Then, the following representation for the solution to (2.48) holds

$$\varphi(x) = \psi_0(x) + \psi_1(x)\frac{\int_G \gamma^T(y)\psi_0(y)dy}{1 - \int_G \gamma^T(y)\psi_1(y)dy}.$$

This relation is a corollary of the general Theorem 2.7 proved below.

Let

$$M[x,y] = K[x,y] - \sum_{k=1}^{n} \delta_k(x)\gamma_k^T(y). \tag{2.51}$$

Introduce $n+1$ auxiliary systems of equations

$$\begin{aligned}\psi_0(x) &= \int M[x,y]\psi_0(y)dy + f(x),\\ \psi_1(x) &= \int M[x,y]\psi_1(y)dy + \delta_1(x),\\ \psi_n(x) &= \int M[x,y]\psi_n(y)dy + \delta_n(x).\end{aligned} \tag{2.52}$$

Denote by A the matrix with elements

$$a_{ij} = \int \gamma_i^T(y)\psi_j(y)dy$$

and denote by b the vector with components

$$b_i = \int \gamma_i^T(y)\psi_0(y)dy, \quad i,j = 1,\dots,n.$$

Theorem 2.6. *Assume that the system (2.52) is uniquely solvable and there exists an operator $(I-A)^{-1}$. Then the solution to (2.48) is represented as*

$$\varphi(x) = \psi_0 + \psi^T(x)J,$$

where the vector J is determined from the system of linear algebraic equations

$$J = AJ + b. \tag{2.53}$$

Proof. Substituting $K[x,y]$ from (2.51) into (2.48) yields

$$\begin{aligned}
\varphi(x) &= \int M[x,y]\varphi(y)dy + \delta_1(x)\int \gamma_1^T(y)\varphi(y)dy + \cdots \\
&\quad \cdots + \delta_n(x)\int \gamma_n^T(y)\varphi(y)dy + f(x) \\
&= \int M[x,y]\varphi(y)dy + \sum_{i=1}^{n}\delta_i(x)J_i + f(x),
\end{aligned} \tag{2.54}$$

where

$$J_i \equiv \int \gamma_i^T(y)\varphi(y)dy = \int \sum_{j=1}^{n}\gamma_i^j(y)\varphi^j(y)dy.$$

Now,

$$\psi_0(x) + \sum_{i=1}^{n}\psi_i(x)J_i = \int M[x,y]\Big\{\psi_0(y) + \sum_{i=1}^{n}\psi_i(y)J_i\Big\}dy + \sum_{i=1}^{n}\delta_i(x)J_i + f(x). \tag{2.55}$$

Comparison of (2.54) and (2.55) shows that

$$\varphi(x) = \psi_0(x) + \sum_{i=1}^{n}\psi_i(x)J_i.$$

Note that

$$J_i = \int \gamma_i^T(y)\varphi(y)dy = \int \gamma_i^T(y)\Big\{\psi_0(y) + \sum_{j=1}^{n}\psi_j(y)J_j\Big\}dy,$$

or

$$J_i = J_1\int \gamma_i^T(y)\varphi_1(y)dy + \cdots + J_n\int \gamma_i^T(y)\psi_n(y)dy + \int \gamma_i^T(y)\psi_0(y)dy,$$

for $i = 1, \ldots, n$, i.e., the vector J satisfies the system (2.53). □

Note that the approach described can be applied also to a system of linear equations $x = Ax + b$, where A is an $m \times m$ matrix.

We introduce a matrix

$$B = A - \alpha_1\beta_1^T - \ldots - \alpha_n\beta_n^T,$$

where $\alpha_1, \dots, \alpha_n$ and $\beta_1, \dots, \beta_n$ are arbitrary column vectors, i.e., the matrix B is obtained from A by subtraction of singular matrices of the following form ($n < m$):

$$\alpha_i \beta_i^T = \begin{pmatrix} \alpha_i^1 \beta_i^1 & \alpha_i^1 \beta_i^2 & \dots & \alpha_i^1 \beta_i^n \\ \alpha_i^2 \beta_i^1 & \alpha_i^2 \beta_i^2 & \dots & \alpha_i^2 \beta_i^n \\ \dots & \dots & \dots & \dots \\ \alpha_i^n \beta_i^1 & \alpha_i^n \beta_i^2 & \dots & \alpha_i^n \beta_i^n \end{pmatrix}.$$

Consider $n+1$ auxiliary linear systems

$$\begin{aligned} x_0 &= Bx_0 + b, \\ x_1 &= Bx_1 + \alpha_1, \\ &\dots \\ x_n &= Bx_n + \alpha_n. \end{aligned}$$

As in the cases described above

$$x = x_0 + \sum_{j=1}^{n} J_j x_j,$$

where $J_1, \dots J_m$ are components of the vector J, which satisfies the equation $J = TJ + t$. Here, T is the matrix with elements $T_{ij} = \beta_i^T x_j$, and t is the vector with components $t_i = \beta_i^T x_0$.

The results presented can be generalized to the case

$$K(x, y) = M(x, y) + \sum_{i=1}^{n} \alpha_i(x) \beta_i(y).$$

Consider $n+1$ auxiliary equations

$$\begin{aligned} \varphi_0(x) &= \int_G M(x, y) \varphi_0(y) dy + f(x), \\ \varphi_i(x) &= \int_G M(x, y) \varphi_i(y) dy + \alpha_i(x), \quad i = 1, \dots, n. \end{aligned}$$

Assuming that these equations are uniquely solvable, we derive, as previously, that

$$\varphi(x) = \varphi_0(x) + \sum_{i=1}^{n} \varphi_i(x) J_i,$$

where $J = (J_1, \dots, J_n)^T$ is determined from

$$J = TJ + b,$$

T being a matrix with elements

$$T_{ij} = \int_G \varphi_j(y)\beta_i(y)dy,$$

$$b_i = \int_G \varphi_0(y)\beta_i(y)dy, \quad i,j = 1,\dots,n,$$

provided that $\det T \neq 0$.

2.5 Integral equation of the first kind

In this section, we consider the equation

$$\int_\Gamma k(x,y)\varphi(y)dy = f(x), \quad x \in \Gamma \tag{2.56}$$

or in the operator form

$$K\varphi = f,$$

where K is a completely continuous, positive, self-adjoint operator: $K : L_2(\Gamma) \to L_2(\Gamma)$. Here, Γ is a $(m-1)$-dimensional hypersurface in $\mathbb{R}^m$ or a bounded domain in $\mathbb{R}^{m-1}$. Assuming that there exists a unique solution to (2.56), we construct a Monte Carlo estimator.

We rewrite (2.56) in the form ($\kappa \neq 0$) (for references see [98])

$$\varphi = (I - \kappa K)\varphi + \kappa f.$$

It is known that

$$\varphi_n \equiv \sum_{i=0}^{n}(I-\kappa K)^i \kappa f \to \varphi \tag{2.57}$$

as $n \to \infty$, if

$$0 < \kappa < 2\|K\|^{-1}.$$

We represent φ_n as a polynomial of K

$$\varphi_n = \kappa \sum_{i=0}^{n} C_{n+1}^{i+1}(-\kappa)^i K^i f.$$

Suppose that it is desired to find the solution to (2.56) at a point $x \in \Gamma$.

We define a Markov chain, starting at $x : Y_0 = x$, with a transitional density, $p(Y_i, Y_{i+1})$, consistent with the kernel $k(Y_i, Y_{i+1})$ of the integral operator K. Following Section 2.1, we can use the adjoint estimator for the iterations:

$$K^i f = \mathbb{E} Q_i^\star f(Y_i), \quad Q_0^\star = 1.$$

Consequently,

$$\xi^\star(n; x) = \kappa \sum_{i=0}^{n} C_{n+1}^{i+1} (-\kappa)^i Q_i^\star f(Y_i) \tag{2.58}$$

is an unbiased estimator for φ_n:

$$\varphi_n(x) = \mathbb{E}\xi^\star(n; x).$$

It is not difficult to estimate the error:

$$||\varphi - \varphi_n||^2 \le [1 - \kappa\lambda_m^{-1}]^{2(n+1)} ||\varphi||^2 + \sum_{k=m+1}^{\infty} (\varphi, f_k)^2, \tag{2.59}$$

where $\lambda_1 \le \lambda_2 \le \dots$ are the characteristic values of K, and $f_1, f_2, \dots$ are the eigenfunctions.

Suppose now that it is desired to calculate a linear functional $I_h = (\varphi, h)$ of the solution φ. Then

$$(\varphi_n, h) = \mathbb{E}\xi_n,$$

where

$$\xi_n^\star = \frac{\xi^\star(n; Y_0) h(Y_0)}{p_0(Y_0)}.$$

We can calculate the variance of this estimator:

$$\mathbb{D}\zeta_n^\star(n, x) = \sum_{i=1}^{n+1} (C_n^i)^2 (-\kappa)^{2i} d_{ii} + 2 \sum_{j>i=1}^{n+1} C_n^j C_n^i (-\kappa)^{i+j} d_{ij},$$

where

$$d_{ij} = \left(K_2^{i-1} (f K^{j-i} f), \frac{h^2}{p_0} \right) - (K^{i-1} f, h)(K^{j-1} f, h),$$

$$K_2 f(x) = \int_\Gamma \frac{k^2(x, y)}{p(x, y)} f(y) dy.$$

From this, we get

$$\mathbb{D}\zeta^\star(n; x) \le \max d_{ij} (1 + \kappa)^{2n}. \tag{2.60}$$

Consider now a different approach based on the iteration of the resolvent operator. We choose κ so that $(I+\kappa K)^{-1}$ exists. Then we can rewrite (2.56) in the form

$$\varphi = (I+\kappa K)^{-1}\varphi + \kappa(I+\kappa K)^{-1}f.$$

For $\kappa > 0$,

$$\varphi_n = \sum_{i=1}^{n+1}[(I+\kappa K)^{-1}]^i \kappa f \to \varphi \tag{2.61}$$

as $n \to \infty$.

We assume now that

$$k(x,y) = O(|x-y|^{\frac{1-m}{2}+\varepsilon})$$

as $|x-y| \to 0$, $\varepsilon > 0$. Then we can construct an unbiased estimator (with a finite variance) for the resolvent operator R, defined by

$$(I+\kappa K)^{-1} = I - \kappa R.$$

The kernel $r(x,y)$ of the operator R satisfies the equation (y is considered as a parameter):

$$r(x,y) = -\kappa K r(x,y) + k(x,y), \quad x \in \Gamma.$$

Hence, for $0 < \kappa < ||K||^{-1}$, we can define an estimator

$$\zeta(x,y) = \sum_{i=0}^{N} Q^{\star}_{i,1} k(X_i, y). \tag{2.62}$$

Here, $\{X_i\}$ is a terminating Markov chain with a transitional density $l(X_{i-1}, X_i)$, $X_i \in \Gamma$, $X_0 = x$; N is the random number of states, and

$$Q^{\star}_{0,1} = 1, \quad Q^{\star}_{i,1} = Q^{\star}_{i-1,1}\left[-\kappa \frac{k(X_{i-1}, X_i)}{l(X_{i-1}, X_i)}\right].$$

Instead of exact values of $r(y_{k-1}, y_k)$, we use their unbiased estimators $\zeta(y_{k-1}, y_k)$. From (2.61), we derive

$$\mathbb{E}\xi^{\star}(n,x) = \varphi_{n-1}(x),$$

where

$$\xi^{\star}(n;x) = n\kappa f(x) + \kappa \sum_{k=1}^{n} C_{n+1}^{k+1}(-\kappa)^k Q^{\star}_{k,2} f(Y_k), \tag{2.63}$$

$$Q^{\star}_{0,2} = 1, \quad Q^{\star}_{k,2} = Q^{\star}_{k-1,2}\zeta(Y_{k-1}, Y_k)/p(Y_{k-1}, Y_k).$$

In the general case, however, it is not possible to construct a random estimator for $r(x,y)$ with a finite variance. Then, we choose $\kappa: \ 0 < \kappa < ||K||^{-1}$ and take N terms in the representation

$$(I + \kappa K)^{-1} = \sum_{k=0}^{\infty} C_{i+k-1}^{k} (-\kappa K)^k .$$

Next, we substitute the resulting polynomial into (2.61) and arrive at the estimator:

$$\xi^*(n;x) = (n+1)\kappa f + \kappa \sum_{k=1}^{N} (-\kappa)^k C_{n+1}^{k+1} Q_k^* f(Y_k). \tag{2.64}$$

It is also possible to use a vector random estimator.

Denote in (2.61) $\chi_i = \varphi_{i-1}$, $i = 1, \dots, n$ and rewrite it in the form

$$\chi_1 = -\kappa K \chi_1 + \kappa f,$$
$$\chi_i = \chi_{i-1} - \kappa K \chi_i, \ \ i = 1, 2, \dots, n,$$

or in a matrix-operator form

$$\chi = \mathcal{K}\chi + F,$$

where $\chi = (\chi_1, \dots \chi_n)^T$.

We choose κ so that the spectral radius $\rho(\kappa K)$ is less than 1. Then, the Neumann series for $\mathcal{K}$ is convergent, and we can use the approximation

$$\varphi_{n-1} = \sum_{i=1}^{n} \chi_i .$$

Hence, we can take

$$\varphi_{n-1}^{(N)} = \sum_{i=1}^{n} \left\{ \sum_{k=0}^{N} \mathcal{K}^k F \right\}_i ,$$

where the subscript 'i' shows that the ith component of the vector is taken. Since n and N are finite, for $N \le n-1$, we get

$$\varphi_{n-1}^{(N)} = \sum_{k=0}^{N} (I - \kappa K)^k \kappa f .$$

This implies that $\varphi_{n-1}^{(N)} = \varphi_N$ in (2.57).

We make now some remarks about the computational cost of the estimators proposed. First of all, this cost depends on the rate of convergence of the Neumann series for the operator $I - \kappa K$. If the set of characteristic numbers of K is bounded, then there exists such $q \in (0,1)$ that

$$||\varphi - \varphi_n|| < cq^n.$$

It is reasonable to adjust the statistical error to the bias of ξ^*. This ensures the consistency when calculating $\varphi(x)$ by averaging over N independent samples of ξ^*. Let

$$\varepsilon = Cq^n = \left(\frac{(1+\kappa)^{2n}}{N}\right)^{1/2}.$$

Then, $n = \ln\varepsilon / \ln q + C_1$,

$$N = C_2 \varepsilon^{2\left\{\frac{\ln(1+\kappa)}{\ln q} - 1\right\}}.$$

If $q = 1 - \alpha$, $\alpha \ll 1$, then the cost of the estimator ξ^* is

$$T = N \cdot n = \text{const}\left(-\frac{1}{\alpha}\ln\varepsilon + C_1\right)\varepsilon^{-2-2\alpha\ln(1+\kappa)}.$$

Assume that $\lambda_m = O(m)$ (e.g. this is true in the case of the single-layer potential for convex Γ). Then, $n = \text{const}/\varepsilon$ and

$$T = O(\varepsilon^{-2}\exp\{\varepsilon^{-1} \cdot C \cdot \ln(1+\kappa)\}).$$

3 Random walk-on-boundary algorithms for the Laplace equation

3.1 Newton potentials and boundary integral equations of the electrostatics

In this chapter we consider classical BVPs of the potential theory in the stationary case. It means that we are concerned with the Laplace equation

$$\Delta u(x) = 0, \quad x \in G,$$

where function u is defined in some region G of the Euclidean space $\mathbb{R}^m$ and has continuous derivatives of at least second order.

Let G be a bounded simple connected domain with a simple connected boundary $\Gamma = \partial G$. We denote $G_1 = \mathbb{R}^m \setminus \overline{G}$, $\overline{G} = G \cup \Gamma$. From here on, m is considered to be greater than or equal to 3, which makes it possible to write down the general expressions of the formulas, but it must be mentioned that the two-dimensional (2D) case can be treated in the same way.

We pass on to the definition of the surface potentials now. Suppose at first that Γ is a smooth Lyapunov surface [34]. It means that

(1) at every point $y \in \Gamma$, there exists the definite normal vector $n(y)$;

(2) this vector, considered as a function of the point y on the surface Γ, is Hölder continuous, that is if $x, y \in \Gamma$, then

$$|n(y) - n(x)| \leq A|x - y|^{\alpha}$$

for some particular constants A and $\alpha \in (0, 1]$;

(3) there exists such constant $d > 0$ that if we consider the ball $S(y, d)$ with the centre at some point $y \in \Gamma$ and the radius equal to d, then the straight lines parallel to $n(y)$ intersect Γ in $S(y, d)$ only once.

Let σ be a standard surface measure on Γ and $\mu(y)$, $y \in \Gamma$ – some continuous function. Then, we can introduce the following functions (see for example [120]): function

$$V(x) = \int_{\Gamma} \frac{2}{(m-2)\sigma_m} |x - y|^{2-m} \mu(y) d\sigma(y) \equiv V[\mu](x), \tag{3.1}$$

which is given the name single-layer potential and function

$$\begin{aligned} W(x) &= \int_{\Gamma} \frac{2}{(m-2)\sigma_m} \frac{\partial}{\partial n(y)} |x - y|^{2-m} \mu(y) d\sigma(y) \\ &= \int_{\Gamma} \frac{2}{\sigma_m} \frac{\cos \varphi_{yx}}{|x - y|^{m-1}} \mu(y) d\sigma(y) \equiv W[\mu](x), \end{aligned} \tag{3.2}$$

which is named double-layer potential. We denote here

$$\sigma_m = \frac{2\pi^{\frac{m}{2}}}{\Gamma(\frac{m}{2})},$$

the surface area of a unit sphere in $\mathbb{R}^m$, $\Gamma(\cdot)$ is the standard $\Gamma-$ function.

Hereafter we shall assume that $n(y)$ is the **internal** (inward pointing) normal vector, φ_{yx} is the angle between vectors $n(y)$ and $x-y$, $\cos\varphi_{yx} = \frac{(n(y),x-y)}{|x-y|}$, where $(.,.)$ is a scalar product of two vectors in $\mathbb{R}^m$. Properties of potentials V and W are thoroughly studied (see, for example, [34, 120]). First of all, $V(x)$ and $W(x)$ are harmonic functions everywhere in $\mathbb{R}^m \setminus \Gamma$ even for a piecewise smooth surface Γ.

Theorem 3.1. *If Γ is a Lyapunov surface and μ is a continuous function, then*
(1) *$V \in C(\mathbb{R}^m)$ (and is continuous on Γ consequently);*
(2) *there exist regular normal derivatives $\left(\frac{\partial V}{\partial n(x)}\right)_+$ and $\left(\frac{\partial V}{\partial n(x)}\right)_-$ on Γ, and*

$$\left(\frac{\partial V}{\partial n(x)}\right)_\pm = \pm\mu(x) + \frac{\partial V}{\partial n(x)} = \pm\mu(x) + \int_\Gamma \frac{2}{\sigma_m} \frac{\cos\varphi_{xy}}{|x-y|^{m-1}}\mu(y)d\sigma(y). \tag{3.3}$$

We remind that function $u(x)$ is said to have regular normal derivative on Γ if, uniformly in all $x \in \Gamma$, there exists

$$\lim_{x'\to x} \frac{\partial u(x')}{\partial n(x)} \equiv \lim_{x'\to x} (\nabla u(x'), n(x)) = \left(\frac{\partial u(x)}{\partial n(x)}\right)_\pm, x' \in \mp n(x),$$

that is $(\cdot)_+$ denotes the limit from the exterior and $(\cdot)_-$ denotes the limit from the interior of the domain G. Note that $\frac{\partial V}{\partial n(x)}$ is named the direct value of the normal derivative of a single-layer potential and it is continuous on Γ.

An analogous theorem is valid for the double-layer potential.

Theorem 3.2. *If Γ is a Lyapunov surface and μ is a continuous function, then*
(1) *$W \in C(\overline{G})$ and $C(\overline{G}_1)$;*
(2) *limiting values of W satisfy the following relation:*

$$(W(x))_\pm = \mp\mu(x) + W(x) = \mp\mu(x) + \int_\Gamma \frac{2}{\sigma_m} \frac{\cos\varphi_{yx}}{|y-x|^{m-1}}\mu(y)d\sigma(y), \tag{3.4}$$

where $W(x)$ is the direct value of a double-layer potential, and this function is continuous on Γ.

These two formulas ((3.3) and (3.4)) will be widely used in this chapter. They serve as a foundation for the integral equations construction and, as a consequence, Monte Carlo estimators for solutions of the basic BVPs for the Laplace equation. In what follows, we pass on to the concrete classes of problems.

3.2 The interior Dirichlet problem and isotropic random walk-on-boundary process

Consider the following problem:

$$\begin{aligned} \Delta u(x) &= 0, \quad x \in G \\ u(y) &= g(y), \quad y \in \Gamma \end{aligned} \tag{3.5}$$

for some continuous function g. It is worth noting, that the last equality means, that the interior limit $(u)_-$ of the function u is equal to g. If the domain G and its boundary Γ satisfy the above-mentioned conditions, then the following statement is valid [120].

Theorem 3.3. *The interior Dirichlet problem (3.5) is solvable for all continuous g, and its solution can be represented in the form of a double-layer potential (3.2).*

So, in the conditions of this theorem, we seek a solution of the problem (3.5) in the form of a double-layer potential W with an unknown density μ. As a consequence of relation (3.4), we come to the integral equation of the second kind for this function:

$$\mu(y) = \lambda \int_{\Gamma} k(y, y')\mu(y')d\sigma(y') + g(y), \quad y \in \Gamma, \tag{3.6}$$

where $\lambda = 1$, and the kernel of the integral operator K of this equation is equal to

$$k(y, y') = -\frac{2}{\sigma_m} \cdot \frac{\cos \varphi_{y'y}}{|y - y'|^{m-1}}. \tag{3.7}$$

On Lyapunov surfaces, this kernel is weakly singular [120]. Therefore, the operator K is compact [47] and, as a consequence of its compactness, all the poles of the resolvent of the integral equation (3.6) coincide with the characteristic values of this equation. The set of these numbers is at most countable, and we can enumerate them in accordance with their moduli: $|\lambda_1| \le |\lambda_2| \le \dots$. To construct a Monte Carlo estimator for the solution of the integral equation (3.6) and hence for the problem (3.5), we have to know the particular spectral properties of the equation (3.6). Simple but effective technique of the Green's formulas provides a possibility to show that [32, 34]

(1) all characteristic values are real numbers, and $|\lambda_i| \ge 1$;
(2) they are simple poles of the resolvent;
(3) $\lambda_1 = -1$ is the characteristic value with the least modulus;
(4) $\lambda = 1$ is not a characteristic value.

The last property means that there exists the unique solution of the integral equation (3.6) for $\lambda = 1$. However, due to the third property, the Neumann series does not converge for an arbitrary function g. This is exactly the case, which has been considered in Chapter 2. Thus, we can follow the procedure that has been written out there.

Note first of all that by setting $u(x) = W(x)$, we represent the solution in the form of an integral functional

$$u(x) = (h_x, \mu),$$

where μ is a solution of the integral equation (3.6) and the weight function is defined as follows:

$$h_x(y) = \frac{2}{\sigma_m} \cdot \frac{\cos\varphi_{yx}}{|x-y|^{m-1}}. \tag{3.8}$$

To begin with, let G be convex. In this case, $h_x > 0$ for all $x \in G$ and $h_x(y_0) = 2p_0(y_0)$ for $x_0 = x$, where

$$p_0(y_0) = \frac{1}{\sigma_m} \cdot \frac{\cos\varphi_{y_0x_0}}{|x_0-y_0|^{m-1}} \tag{3.9}$$

is the probability density function corresponding to the distribution of a point y on Γ uniform in a solid angle with its vertex at point x_0. This fact follows from a simple geometric consideration (see [34] as an example) that $p_0(y)d\sigma(y) = d\omega(y)$ is an elementary solid angle. In the full analogy, we have that

$$p(y_i, y_{i+1}) = \frac{2}{\sigma_m} \cdot \frac{\cos\varphi_{y_{i+1}y_i}}{|y_i-y_{i+1}|^{m-1}} \tag{3.10}$$

is the probability density function of the uniform distribution of a point y_{i+1} on Γ in a solid angle with its vertex at a point y_i, which is also on Γ.

Definition 3.1. Markov chain $\{y_0, y_1, \ldots\}$ of points y_i on Γ is named the process of isotropic random walk on boundary (WB-process) if its initial point is distributed with the probability density $p_0(y_0)$ and its transitional density is $p(y_i, y_{i+1})$.

To construct a trajectory of an isotropic WB-process, we choose some point x in the interior of G and then sample a ray with a random uniformly distributed direction coming out from this point. The unique in the case of convex Γ intersection of this ray with the boundary gives us the starting point y_0 of the Markov chain. The analogous procedure is used to find the next point y_{i+1} when the previous y_i is known. The only difference is that in this case, the direction $y_{i+1} - y_i$ is uniformly distributed in a half-space determined by the tangent plane at the point y_i. It should be noted that we do not need to know the real positioning of the tangent plane since an isotropic distribution does not depend on the coordinate system. As a consequence, we can proceed as follows. Choose randomly a straight line (not a ray) with an isotropic direction in a fixed coordinate system. This line, as opposed to a ray, always intersects boundary Γ in the unique not equal to y_i point, which we choose as y_{i+1}.

Now we turn to definition of an estimator for the solution of BVP (3.5). It follows from (3.7) to (3.10) that the kernel of the integral equation (3.6) $k(y, y')$ is consistent with the transitional density $p(y, y')$ of the isotropic WB-process, and more than that,

$|k(y, y')| = p(y, y')$. So it is natural to use adjoint estimators (Chapter 2) for iterations of the integral operator K and hence for a solution of BVP (3.5). As a result, we have $Q_0^\star = 2$, $Q_i^\star = 2 \cdot (-1)^i$, and

$$\xi^\star(x; n) = \sum_{i=0}^{n} 2\, l_i^{(n)} (-1)^i g(y_i) \tag{3.11}$$

is $\delta(x; n)$-biased estimator for $u(x)$, where $|\delta(x; n)| \le \text{const} \cdot q^n$, and $q < 1$; $l_i^{(n)}$ are defined by the concrete method of analytical continuation of the resolvent (Chapter 2).

In this particular case, the characteristic set $\chi(K)$ is well-defined, and we can use comparatively simple domains D in the complex plane λ with the required properties. As an example, we put $D = \{\lambda : \Re\lambda > -1\}$, i.e., the half-plane, and take the function $\lambda = 2\eta/(1 - \eta)$, which maps the unit disk onto D. In this case,
$\eta^* = 1/3$, $q = \eta^*/\eta_0$, $\eta_0 = \min\left|\frac{\lambda_i}{\lambda_i+2}\right|$
and

$$l_i^{(n)} = \sum_{k=i}^{n} 2^i (1/3)^k C_{k-1}^{i-1}.$$

Another appropriate function is $\lambda = 4\eta/(1 - \eta)^2$. It maps the unit disk onto the whole plane, with the ray of real numbers less than or equal to -1 excluded. Here,
$\eta^* = 3 - 2\sqrt{2}$, $q = \eta^*/\eta_0$, $\eta_0 = \min\left|1 + \frac{2}{\lambda_i} - \frac{2}{\lambda_i}\sqrt{1 + \lambda_i}\right|$
and

$$l_i^{(n)} = \sum_{k=i}^{n} 4^i (3 - 2\sqrt{2})^k C_{i+k-1}^{2i-1}.$$

Remember now that $\lambda_1 = -1$ is a simple pole of the resolvent, R, and $|\lambda_2| > 1$. Thus,

$$R'(\lambda) = (\lambda + 1) R_\lambda$$

is an analytical function of λ in the interior of the disk $\{\lambda : |\lambda| < |\lambda_2|\}$. As a consequence, this function can be expanded in the convergent operator series

$$R'(\lambda) = K + \lambda(K + K^2) + \lambda^2(K^2 + K^3) + \dots,$$

and since $R_1 = \frac{1}{2} R'(1)$, we can proceed as follows. Having taken a finite number of terms, we can evaluate the remainder using the fact that the series for R' converges at the speed of a geometric progression with multiplier $q \le \frac{1}{|\lambda_2|}$. So we have

$$\mu(y) = \sum_{i=0}^{n-1} K^i g(y) + \frac{1}{2} K^n g(y) + \varepsilon(n; y),$$

and the estimator based on this representation is (3.11), where

$$l_i^{(n)} = 1, \quad i = 1, \dots, n-1,$$
$$l_n^{(n)} = \frac{1}{2}.$$

In Chapter 2, we showed that the resolvent analytical continuation and other methods of rearrangement of the original integral equation provide a possibility to construct new iterative algorithms for computing the solution. Often, such iterative procedure may be considered as a summation of Neumann series for a different integral equation. In particular, linear fractional mapping $\lambda = \frac{\alpha\eta}{1-\beta\eta}$ gives birth to a class of alternative representations

$$\mu = \sum_{i=0}^{\infty} K_1^i g_1,$$

where

$$K_1 = \frac{\beta}{\alpha+\beta} I + \frac{\alpha}{\alpha+\beta} K,$$
$$g_1 = \frac{\alpha}{\alpha+\beta} g.$$

Note that we cannot use the standard Monte Carlo procedure of constructing an estimator based on simulating trajectories of a terminating Markov chain. The reason is that the Neumann series for $\overline{K}_1$ does not converge. So we take, as usual, a finite number of terms, n. This number is determined by the required accuracy, i.e.,

$$||\mu - \sum_{i=0}^{n} K_1^i g_1|| \le \varepsilon.$$

In our particular case, we take $\beta = 1$, $0 < \alpha \le 2$ and construct a special Markov chain with a transitional density, which is chosen in accordance with the kernel of the integral operator K_1. The initial point of this chain is sampled according to the density p_0 from (3.9). Next, if a point y_i is defined, we randomize the choice of the point y_{i+1} in the following way. We set $y_{i+1} = y_i$ with probability $1/(1+\alpha)$, and with probability $\alpha/(1+\alpha)$, we choose y_{i+1} uniformly distributed in the solid angle of view from the point y_i. Thus, we have the following estimator:

$$\xi_\alpha^*(x; n) = \frac{\alpha}{1+\alpha} \sum_{i=0}^{n} Q_i^* g(y_i), \tag{3.12}$$

where $Q_0^* = 2$, $Q_{i+1}^* = Q_i^* \cdot b(y_i, y_{i+1})$, and

$$b(y_i, y_{i+1}) = \begin{cases} 1, & \text{with probability} \quad 1/(1+\alpha) \\ -1, & \text{with probability} \quad \alpha/(1+\alpha) \end{cases}.$$

It is worth noting that multiplication of the resolvent by the term $1+\lambda$ leads to some iterative method too, namely,

$$\mu = \frac{1}{2}g + \sum_{i=0}^{\infty} K^i g_1,$$

where $g_1 = \frac{1}{2}(Kg+g)$, and we have to sum the Neumann series for the original operator K. These series converge here, since g_1 is orthogonal to μ_0, the Robin's potential density. It means that $(\mu_0, g_1) = 0$, and $\mu_0 = -K^*\mu_0$ is the characteristic function (and the eigenfunction in this case), corresponding to $\lambda_1 = -1$, the characteristic value of the adjoint integral operator K^*. Lacking the absolute convergence of the Neumann series, we have to take a finite number of terms and construct an estimator based on trajectories of an isotropic WB-process. The adjoint estimators for iterations of the integral operator K are natural to be used here. Therefore, we have to know values of the function g_1 at the points of WB-process. The double randomization principle provides a possibility to use unbiased conditionally independent estimators instead of exact values of $g_1(y_i)$. As a result, we come to the following expression:

$$\xi_0^*(x;n) = \frac{1}{2}Q_0^* g(y_0) + \frac{1}{2}\sum_{i=0}^{n} Q_i^* \left(g(y_i) + \frac{k(y_i, Y_i)}{p(y_i, Y_i)} g(Y_i)\right), \tag{3.13}$$

where Y_i is sampled isotropically in accordance with the density p and is independent of the point y_{i+1}. We have $Q_i^* = 2\cdot(-1)^i$, since we consider the domain G to be convex here.

Recall now that everywhere in this chapter, we have supposed that the point $x \in G$, at which the solution of the BVP is computed, coincides with the point x_0, which is involved in the definition of the density function $p_0(y_0)$ (3.9). However, such requirement is not essential. For estimate to have finite mathematical expectation, it is sufficient to suppose that the ratio

$$Q_0^*(x, y_0; x_0) = \frac{h_x(y_0)}{p_0(y_0)} = 2\frac{\cos\varphi_{y_0 x}\cdot |x_0 - y_0|^{m-1}}{\cos\varphi_{y_0 x_0}\cdot |x - y_0|^{m-1}}$$

is bounded, or even less, that Q_0^* is finite with probability one. As a consequence, solution u can be evaluated at several points $x^{(j)} \in G$ simultaneously. The only condition that should be satisfied is these points must be separated from the boundary (it means that a distance between x_j and Γ is uniformly in j bounded away from zero). Then, $Q_0^*(x^{(j)}, y_0; x_0)$ is uniformly bounded. The adjoint estimator that can be used here differs from (3.11) in its initial weight only:

$$\xi^*(x^{(j)};n) = \sum_{i=0}^{n} Q_0^*(x^{(j)}, y_0; x_0) l_i^{(n)} (-1)^i g(y_i). \tag{3.14}$$

Now we turn our attention to the problem of evaluating the derivatives of the solution. This task is often more important than solving the BVP itself. Let x be an interior point in G. Analyticity of a double-layer potential $W(x)$ at this point provides a possibility to differentiate this function under the integral sign. So we have

$$\frac{\partial u}{\partial x_k}(x) = \left(\frac{\partial h_x}{\partial x_k}, \mu\right), \quad k = 1, \ldots, m,$$

where

$$\frac{\partial h_x}{\partial x_k}(y) = \frac{2}{\sigma_m} \cdot \frac{1}{|x-y|^m}\left(n_k(y) - m \cdot \cos\varphi_{yx} \cdot \frac{x_k - y_k}{|x-y|}\right),$$

$x_k, y_k, n_k(y)$ are coordinates of corresponding vectors $x, y, n(y)$. It is clear that if we take a finite number of points $x^{(j)}$ in the interior of a convex domain G separated from its boundary Γ, the same estimator (3.12) can be used for computing not only the solution at these points but also all its first derivatives. The difference is contained in the weight $Q_0^\star$ only, i.e.,

$$Q_0^\star(x^{(j)}, y_0; x_0) = \frac{\partial h_x^{(j)}}{\partial x_k}(y_0) \cdot (p_0(y_0))^{-1} \tag{3.15}$$

and

$$Q_0^\star(x_0, y_0; x_0) = \frac{2}{|x_0 - y_0|}\left(\frac{n_k(y_0)}{\cos\varphi_{y_0x_0}} - m\frac{x_{0k} - y_{0k}}{|x_0 - y_0|}\right)$$

in the case $x = x_0$.

It has been assumed so far that domain G is convex. Let us pass on to a more general case. The first consequence of a non-convexity is that function $p(y_i, y_{i+1})$ defined in (3.10) is not positive for all points y_{i+1}, and therefore, it cannot be used as a transitional density. From the geometric point of view, it means that a ray starting at the point y_i (or at the point x_0) can intersect Γ at several points, and in each of them, $\cos\varphi_{y_{i+1}y_i}$ has its own sign.

The inherent relation between the kernel k and the isotropic random distribution of the next point of the Markov chain, the convenience of such random choice, leads to the desire of using the same p from (3.10) for constructing WB-process in a non-convex case. One of the possible ways to do this is following. Denote by $q(y_i, y_{i+1}) \geq 1$ the number of intersections of an isotropically distributed straight line through the points y_i, y_{i+1}, excluding y_i. Then, with probability $1/q$, we choose randomly one of these intersections and set it to be y_{i+1}. The transitional density that corresponds to such choice of the next point in the Markov chain is

$$p(y_i, y_{i+1}) = \frac{2}{\sigma_m} \cdot \frac{|\cos\varphi_{y_{i+1}y_i}|}{q(y_i, y_{i+1}) \cdot |y_i - y_{i+1}|^{m-1}}. \tag{3.16}$$

In the full analogy, the distribution density of the initial point is equal to

$$p_0(y_0) = \frac{1}{\sigma_m} \cdot \frac{|\cos\varphi_{y_0x_0}|}{q(x_0, y_0) \cdot |x_0 - y_0|^{m-1}}. \tag{3.17}$$

A Markov chain constructed on a non-convex Γ in accordance with the densities p_0 and p that are defined by formulas (3.17), (3.16) is also named an isotropic walk-on-boundary process.

Having transformed the distribution densities that define WB-process, we inevitably come to changes in the form of estimators already constructed for the convex case. So we have

$$\xi^*(x;n) = \sum_{i=0}^{n} l_i^{(n)} Q_0^*(-1)^i T_i g(y_i), \tag{3.18}$$

where

$$T_j = \prod_{j=0}^{i-1} q(y_j, y_{j+1}) \cdot \text{ sign}\,(\cos\varphi_{y_{j+1}y_j}), \quad T_0 = 1. \tag{3.19}$$

In the full analogy with the convex case, the same trajectories of a WB-process can be used for simultaneous computation of a solution and its derivatives. We take the same estimator (3.17) changing the initial weight Q_0^* only. Setting

$$Q_0^* = \frac{h_x(y_0)}{p_0(y_0)} = \frac{2\cos\varphi_{y_0x}\cdot|x_0-y_0|^{m-1}\cdot q(x_0,y_0)}{|\cos\varphi_{y_0x_0}|\cdot|x-y_0|^{m-1}}, \tag{3.20}$$

we obtain the estimator for a solution $u(x)$. In particular, for $x = x_0$,

$$Q_0^* = 2q(x_0, y_0)\cdot \text{sign}(\cos\varphi_{y_0x_0}).$$

It is essential to note that in non-convex domains, the expression in (3.20) can be unbounded for $x \neq x_0$. However, it does not mean that we cannot use WB-process for simultaneous computation of a solution $u(x)$ at several different points. It is sufficient to change the distribution density p_0 in such a way that $p_0(y) \geq c > 0$ for nearly all $y \in \Gamma$. The same method of changing the initial distribution works in the case when we want to evaluate the derivatives of a solution and weights (3.15) are unbounded due to a non-convexity.

3.3 Solution of the Neumann problem

Another classical well-posed problem is the Neumann BVP for the Laplace equation. It means that

$$\begin{aligned} &\Delta u(x) = 0, \quad x \in G_1(\text{ or } x \in G), \\ &\frac{\partial u}{\partial n}(y) = g^*(y), \quad y \in \Gamma, \end{aligned} \tag{3.21}$$

and $u(x) \to 0$ for $|x| \to \infty$ if we consider the exterior problem. The following theorem [120] is valid here.

Theorem 3.4. *The exterior Neumann problem (3.21) is solvable for all continuous g^* and its solution can be represented in the form of a single-layer potential (3.1).*

As a consequence of this theorem, we can seek for a solution of the Neumann problem in the exterior domain G_1 in the form of a single-layer potential with an unknown density μ^*. This potential satisfies the Laplace equation at any point $x \in G_1$. Combination of the boundary condition $\left(\frac{\partial V}{\partial n}\right)_+ = g^*$ and the formula (3.3) for the potential normal derivative makes it possible to write down the following integral equation of the second kind:

$$\mu^*(y) = \lambda \int_\Gamma k^*(y, y')\mu^*(y')d\sigma(y') + g^*(y), \tag{3.22}$$

or in the operator form

$$\mu^* = \lambda K^* \mu^* + g^*,$$

where $\lambda = 1$, and K^* is the integral operator adjoint to the integral operator K from equation (3.6). Its kernel is

$$k^*(y, y') = k(y', y) = -\frac{2}{\sigma_m} \cdot \frac{\cos \varphi_{yy'}}{|y - y'|^{m-1}}. \tag{3.23}$$

Spectral properties of the operator K are known. It means that we can easily obtain such properties for K^* also. The characteristic set $\chi(K)$ lies on the real axis of a complex plane, and thus $\chi(K^*) = \bar{\chi}(K) = \chi(K)$, where $\bar{\lambda}$ stands for a complex number conjugate to λ. It means that characteristic values of operators K and K^* coincide, and we come to the possibility of using for integral equation (3.22) the same methods as for the integral equation (3.6).

It is obvious that representation of a solution in the form of a single-layer potential can be considered as an integral functional

$$u(x) = (\mu^*, h_x^*),$$

where

$$h_x^*(y) = \frac{2}{(m-2)\sigma_m} \cdot \frac{1}{|x - y|^{m-2}}.$$

This means that we find ourselves in a standard situation considered in Chapter 2 and, for the particular integral equation, in Section 3.2. As a consequence of (3.23), we come to a conclusion that an isotropic WB-process can be used here. Its transitional density $p(y_i, y_{i+1})$ is consistent with the kernel $k^*(y_{i+1}, y_i)$ and so it is natural to use a direct Monte Carlo estimator.

Let us begin with a convex case. In this instance

$$|k^*(y_{i+1}, y_i)| = p(y_i, y_{i+1}), \quad Q_i = (-1) \cdot Q_{i-1},$$

and so the $\delta(x;n)$-biased estimator for a solution of the Neumann problem (3.21) looks like as follows:

$$\xi(x;n) = \sum_{i=0}^{n} l_i^{(n)} Q_0 (-1)^i h_x^*(y_i). \tag{3.24}$$

It has been mentioned already that the same transformations of an integral equation that have been used in the case of the interior Dirichlet problem to obtain a convergent series is appropriate for equation (3.22) also. As a result, the same analytical continuation methods and, hence, coefficients $l_i^{(n)}$ can be used here. The main difference between the estimators (3.11) and (3.24) lies in the initial weight $Q_0 = g^*(y_0)/p_0(y_0)$. Recall now that the point x where the solution is estimated lies outside of G, and so there is no natural choice of a point x_0 inside G that can be used for the density p_0 definition (3.9). So the distribution of the initial point of WB-process is arbitrary to some extent. In a convex case, one of the options could be the isotropic density function (3.9), where $x_0 \in G$ is a suitably chosen point.

It is always desired that an estimator is bounded with probability 1. In that instance, such condition is satisfied if the point x is separated from the boundary, Γ. Here, $x \in G_1$ is the point, at which we want to find a solution. It is obvious that a solution of the exterior Neumann problem and its derivatives can be computed simultaneously using the same trajectories of a WB-process. The last assertion is true since a single-layer potential is an analytical function at the inner points of the domain. So

$$\xi_k(x;n) = \sum_{i=0}^{n} l_i^{(n)} Q_0 (-1)^i \frac{2}{\sigma_m} \cdot \frac{y_{ik} - x_k}{|y_i - x|^m}$$

is an estimator for the derivative $\frac{\partial u}{\partial x_k}(x)$, where a point $x = (x_1, x_2, \ldots, x_m)$ is separated from the boundary Γ. This expression is obtained from the formula (3.24), in which the function $h_x^*(y_i)$ is substituted by its derivative $\frac{\partial}{\partial x_k} h_x^*(y_i)$.

In the full analogy with (3.12), (3.13), to construct a different estimator for the solution of BVP (3.21), we can use representations of μ^* in the form of a Neumann series for a transformed integral equation. As a result, we have

$$\mu^* = \sum_{i=0}^{\infty} (K_1^*)^i g_1^*,$$

where

$$K_1^* = \frac{\beta}{\alpha+\beta} I + \frac{\alpha}{\alpha+\beta} K^*,$$
$$g_1^* = \frac{\alpha}{\alpha+\beta} g^*.$$

To construct an estimator based on this representation, we proceed as follows. Let us take such a finite number of terms in the series that is sufficient to attain the desired accuracy. To do this, we use the information on the speed of convergence, which is determined by the linear fractional mapping $\lambda = \frac{\alpha\eta}{1-\beta\eta}$. Next we have

$$u(x) = (\mu^*, h_x^*) = \Big(\sum_{i=0}^{n} (K_1^*)^i g_1^*, h_x^*\Big) + \delta(x;n)$$
$$= \sum_{i=0}^{n} (g_1^*, K_1^i h_x^*) + \delta(x;n).$$

So we see that the same Markov chain that has been defined in the process of the estimator (3.12) construction can be used in this case. Finally, we come to the following estimator (here $\beta = 1$, $0 < \alpha \le 2$):

$$\xi_\alpha(x;n) = \frac{\alpha}{1+\alpha} \sum_{i=0}^{n} Q_i h_x^*(y_i),$$

where

$$Q_0 = \frac{g^*(y_0)}{p_0(y_0)}, \quad Q_{i+1} = Q_i b(y_i, y_{i+1})$$

for the same random variables b that were defined in (3.12).

Consider now the method based on multiplication of the resolvent by the term $1+\lambda$. In the full analogy with the case of integral operator K, we come to the following convergent series for the operator K^*:

$$\mu^* = \frac{1}{2} g^* + \sum_{i=0}^{\infty} (K^*)^i g_1^*,$$

where $g_1^* = (g^* + K^* g^*)/2$ is orthogonal to μ_1, the characteristic function of the operator $(K^*)^* = K$ that corresponds to $\lambda_1 = -1$. Here, μ_1 is constant.

Taking a finite number of terms in this series, we have

$$u(x) = (\mu^*, h_x^*)$$
$$= \frac{1}{2}(g^*, h_x^*) + \Big(\sum_{i=0}^{n} (K^*)^i g_1^*, h_x^*\Big) + \delta(x;n)$$

$$= \frac{1}{2}(g^*, h_x^*) + \sum_{i=0}^{n} (g_1^*, K^i h_x^*) + \delta(x; n)$$

$$= \frac{1}{2}(g^*, h_x^*) + \frac{1}{2}\sum_{i=0}^{n} (g^*, K^i (h_x^* + K h_x^*)) + \delta(x; n).$$

It is essential to note that the last transformation is valid for finite sum only.

Using this expression we construct an estimator, which is based on simulating trajectories of an isotropic WB-process. In the full analogy with (3.13), the double randomization principle gives us a possibility to use unbiased conditionally independent estimators instead of exact values of Kh_x^*:

$$\xi_0(x; n) = \frac{1}{2} Q_0 h_x^*(y_0) + \frac{1}{2}\sum_{i=0}^{n} Q_i \left(h_x^*(y_i) + \frac{k(y_i, Y_i)}{p(y_i, Y_i)} h_x^*(Y_i) \right).$$

Here, Y_i is isotropically distributed with density p and is independent of the point y_{i+1}. In a convex case, we have

$$Q_i = (-1)^i Q_0, \quad Q_0 = \frac{g^*(y_0)}{p_0(y_0)}, \quad \frac{k(y_i, Y_i)}{p(y_i, Y_i)} = -1.$$

We now turn our attention to a non-convex case. It is obvious that the same transformation of WB-process that has been used for the interior Dirichlet problem is applicable for the exterior Neumann problem also. So we can write out the basic estimator for $u(x)$, a solution of (3.21):

$$\xi(x; n) = \sum_{i=0}^{n} l_i^{(n)} Q_0 (-1)^i T_i h_x^*(y_i), \tag{3.25}$$

where $\{y_i\}$ is an isotropic WB-process with the transitional density (3.16) and an initial density p_0, which is consistent with the boundary function g^*, $Q_0 = \frac{g^*(y_0)}{p_0(y_0)}$ and coefficients T_i are defined in (3.19).

Let us consider the interior Neumann problem now. It is well known that for arbitrary boundary values of the normal derivative, a solution of (3.21) for $x \in G$ does not necessarily exist. Namely, the following proposition is valid [120].

Theorem 3.5. *The interior Neumann problem is solvable for all continuous functions g^* satisfying the orthogonality condition*

$$\int_{\Gamma} g^*(y) d\sigma(y) = 0. \tag{3.26}$$

Its solution can be represented in the form of a single-layer potential, and it is unique up to addition of an arbitrary constant function.

Thus, we consider BVP (3.21) in the interior of a domain G and seek its solution in the form of a single-layer potential (3.1) with an unknown density μ^*. In that instance, the boundary condition means that $\left(\frac{\partial V}{\partial n}\right)_- = g^*$ and so we can use Theorem 3.1 to derive an integral equation. From (3.3), we get

$$\mu^*(y) = \lambda \int_\Gamma \frac{2}{\sigma_m} \cdot \frac{\cos \varphi_{yy'}}{|y-y'|^{m-1}} \mu^*(y') d\sigma(y') - g^*(y) \tag{3.27}$$

or

$$\mu^* = \lambda(-K^*)\mu^* - g^*, \quad \lambda = 1.$$

We see that the integral operator here differs from the integral operator K^*, considered in the case of the exterior Neumann problem, only in its sign. From here, we have that $\lambda_1 = 1$ is a characteristic value for the integral equation (3.27). The Fredholm theorems give some conditions of solvability of the equation (3.27) (see [47], for example). A solution exists and is unique if and only if the free term $-g^*$ of the equation (3.27) is orthogonal to all characteristic functions of the integral operator $-K$ corresponding to the characteristic value $\lambda_1 = 1$. It is well known that there exists only one such function and it is constant on Γ. So we see that (3.26) provides all necessary conditions of solvability and a general solution of (3.27) can be written out in the form

$$\mu^* = \mu_0^* + c\mu_0, \tag{3.28}$$

where c is an arbitrary constant, μ_0 is the Robin's potential density and μ_0^* is a particular solution of (3.27). It will be sought for in the form of a sum of the Neumann series for the equation (3.27). From (3.28), we get

$$u(x) = \int_\Gamma \frac{2}{\sigma_m} \cdot \frac{\mu_0^*(y) + c\mu_0(y)}{|x-y|^{m-1}} d\sigma(y) = V_0^*(x) + \text{const},$$

since the Robin potential

$$V_0(x) = \int_\Gamma \frac{\mu_0(y)}{|x-y|^{m-1}} d\sigma(y)$$

is constant in G. This property can be easily obtained from the definition of μ_0 (it means that V_0 is a solution of (3.21) with $g^* = 0$) and the Green's formulas [120]. We set $V_0(x)$ equal to 1.

It is obvious that the Neumann series does not absolutely converge here. Hence, to construct an estimator, we will use a finite number of iterations of the integral operator $-K^*$. So we have

$$
\begin{aligned}
V_0^{\star}(x) &= \Big(\sum_{i=0}^{\infty}(-K^{\star})^i(-g^{\star}), h_x^{\star}\Big) \\
&= \Big(\sum_{i=0}^{n}(-K^{\star})^i(-g^{\star}), h_x^{\star}\Big) + \delta(x;n) \\
&= -\Big(g^{\star}, \sum_{i=0}^{n}(-K)^i h_x^{\star}\Big) + \delta(x;n),
\end{aligned}
\tag{3.29}
$$

where $|\delta(x;n)| \le \text{const}\Big(\frac{1}{|\lambda_2|}\Big)^n$ and λ_2 is the second characteristic value of $-K^{\star}$. In a convex case, the kernel of the integral operator $-K$ is equal to the transition density p from (3.10). So, to construct an estimator for $V_0^{\star}(x)$, it is natural to use an isotropic WB-process. It is essential to note that to make the estimator effective, we have to take an advantage of the orthogonality condition (3.26).

Divide Γ into two parts $\Gamma = \Gamma_+ \cup \Gamma_-$, where

$$
\begin{aligned}
\Gamma_+ &= \{y \in \Gamma : g^{\star}(y) \ge 0\}, \\
\Gamma_- &= \{y \in \Gamma : g^{\star}(y) < 0\},
\end{aligned}
$$

and let $p_1(y), p_2(y)$ be density functions on Γ_+, Γ_-, respectively. Next, we sample two random points y_0^1 and y_0^2 distributed on Γ_+, Γ_- with corresponding densities p_1, p_2 and construct two trajectories of an isotropic WB-process starting from these points. This procedure of branching, which is generally called stratified sampling [23] or, for a special case, antithetic variates [36], leads to an estimator with less variance. So we have

$$
\xi(x;n) = -\sum_{i=0}^{n} \frac{2}{(m-2)\sigma_m}\left[\frac{g^{\star}(y_0^1)}{p_1(y_0^1)} T_i^1 |x - y_i^1|^{2-m} + \frac{g^{\star}(y_0^2)}{p_2(y_0^2)} T_i^2 |x - y_i^2|^{2-m}\right], \tag{3.30}
$$

where upper indices indicate that T_i^k are computed on the corresponding WB-process $\{y_0^k, y_1^k, \dots, y_n^k\}$, $k = 1, 2$. In a convex case, all $T_i^k = 1$, and we can simplify estimator (3.30) by sampling, as an example, y_0^1, y_0^2 uniformly on Γ_+ and Γ_-, respectively. In this case, $p_1 = p_2 = 2g_0^{-1}$, where $g_0 = \int_{\Gamma} |g^{\star}(y)| d\sigma(y)$, and we arrive at a simplified form of the estimate:

$$
\xi(x;n) = -\sum_{i=0}^{n} \frac{g_0}{(m-2)\sigma_m}\left[|x - y_i^1|^{2-m} - |x - y_i^2|^{2-m}\right].
$$

WB-process trajectories do not depend on the point $x \in G$. As a consequence, we can compute a solution of the interior Neumann problem and its derivatives simultaneously at several points separated from the boundary. The last condition is sufficient for the estimator to have values bounded with probability 1. The same is valid for the derivatives' estimators:

$$\xi_k(x;n) = \sum_{i=0}^{n} \frac{2}{\sigma_m} \left[\frac{g^\star(y_0^1)}{p_1(y_0^1)} T_i^1 \frac{x_k - y_{ik}^1}{|x - y_i^1|^m} + \frac{g^\star(y_0^2)}{p_2(y_0^2)} T_i^2 \frac{x_k - y_{ik}^2}{|x - y_i^1|^m} \right],$$

which can be obtained by the direct differentiation of (3.30).

So far we have supposed that we estimate a solution and its derivatives at a point x, which is separated from the boundary. In that case, kernels of potentials $W(x)$, $V(x)$ and their derivatives are uniformly bounded. The situation becomes essentially different when x tends to Γ. Some special methods for constructing effective estimators for the solution values at the points near the boundary are given in Section 3.7.

3.4 Random estimators for the exterior Dirichlet problem

In this section, we consider the Dirichlet problem in G_1, which is exterior to a bounded domain G. It means that we want to find a harmonic function u:

$$\Delta u(x) = 0, \quad x \in G_1, \tag{3.31}$$

which satisfies the boundary condition

$$u(y) = g(y), \quad y \in \Gamma,$$

and the additional requirement, $u(x) \to 0$ for $|x| \to \infty$.

Suppose first that $u(x)$ is represented as a double-layer potential $W[\mu](x)$ with an unknown density μ. The relation (3.4) and the boundary condition $(u)_+ = g$ give rise to the integral equation

$$\mu(y) = \lambda \int_\Gamma \frac{2}{\sigma_m} \cdot \frac{\cos \varphi_{y'y}}{|y - y'|^{m-1}} \mu(y') d\sigma(y') - g(y), \tag{3.32}$$

or

$$\mu = \lambda(-K)\mu - g, \quad \lambda = 1.$$

For an arbitrary g, solution of this equation does not necessarily exist, since $\lambda_1 = 1$ is the characteristic value of the integral operator $-K$. With the Fredholm theorems, we come to a conclusion that the integral equation (3.32) is solvable if and only if

$$(\mu_0, g) = 0, \tag{3.33}$$

where μ_0 is the Robin's potential density. In that case,

$$\mu = \mu_{(0)} + C,$$

where $\mu_{(0)}$ is some particular solution of (3.21) (the sum of the Neumann series of (3.32), for example) and c, which is constant on Γ, is the unique eigenfunction of $-K$ corresponding to $\lambda_1 = 1$. As a consequence, we come to the following [120] statement.

Proposition 3.2. *The exterior Dirichlet problem has a unique solution in the form of a double-layer potential for all continuous functions g that satisfy the orthogonality condition (3.33).*

The uniqueness here follows from the Gauss formulas for a double-layer potential with constant density.
In particular, $\int\limits_\Gamma h_x(y)d\sigma(y) = 0$ when $x \in G_1$ (Günter, 1934 [34]).

In contrast to the analogous condition (3.26) for the case of the interior Neumann problem, the orthogonality condition (3.33) is not natural for the BVP considered here. It originates from the representation of a solution in the form of a double-layer potential. This representation allows us to find only such solutions that tend asymptotically to zero as $O(|x|^{1-m})$ when $|x| \to \infty$. However, in the original setting of the BVP, speed of convergence is not prescribed. It is well known that harmonic functions tend to zero as $O(|x|^{2-m})$ when $|x| \to \infty$. Therefore, if the condition (3.33) is not satisfied, it means only that there is no such solution to the problem that can be represented in the form of a double-layer potential.

Without loss of generality, we may suppose that $0 \in G$. We seek a solution of the exterior Dirichlet problem in the form of a sum of a double-layer potential and a Newton potential $\alpha|x|^{2-m}$ of an unknown charge α positioned at the point $x = 0$:

$$u(x) = W[\mu](x) + \frac{\alpha}{|x|^{m-2}}. \tag{3.34}$$

Taking into consideration (3.4) and the boundary condition, we come to the integral equation

$$\mu(y) = (-K)\mu(y) + f(y),$$

where

$$f(y) = \frac{\alpha}{|y|^{m-2}} - g(y).$$

This equation differs from (3.32) in its free term only. So the same solvability condition must be satisfied, and here it takes the following form:

$$\int\limits_\Gamma \left(\frac{\alpha}{|y|^{m-2}} - g(y)\right)\mu_0(y)d\sigma(y) = 0 \tag{3.35}$$

or

$$\alpha V_0(0) - \int\limits_\Gamma g(y)\mu_0(y)d\sigma(y) = 0,$$

where V_0 is the Robin potential, which is equal to 1 since $0 \in G$. So we have

$$\alpha = \int_{\Gamma} g(y)\mu_0(y)d\sigma(y), \tag{3.36}$$

and the following theorem is valid.

Theorem 3.6. *The exterior Dirichlet problem is solvable for all continuous functions g, and its solution can be represented as a sum of a double-layer potential and the potential*

$$\frac{1}{|x|^{m-2}} \int_{\Gamma} g(y)\mu_0(y)d\sigma(y).$$

Note that $\alpha = 0$ means the condition (3.33) is satisfied. So, the representation $u(x) = W[\mu](x)$ is a special case of a more general situation considered in this theorem.

So we have the convergent Neumann series

$$\mu(y) = \sum_{i=0}^{\infty} (-K)^i f(y),$$

since orthogonality condition (3.35) is satisfied. Lacking the absolute convergence, we have to take a finite number of terms to construct an estimator. As a consequence, we get

$$W(x) = (h_x, \mu) = \left(h_x, \sum_{i=0}^{n} (-K)^i f\right) + \delta(x; n),$$

where $|\delta(x;n)| \leq \text{const}\left(\frac{1}{|\lambda_2|}\right)^n$.

In the full analogy with the interior Neumann problem, we have $\int_{\Gamma} h_x(y)d\sigma(y) = 0$ (see (3.29), (3.26)), and so we process in the same way. Divide Γ into two parts

$$\Gamma = \Gamma_+ \cup \Gamma_-,$$

where

$$\Gamma_+ = \{y \in \Gamma : h_x(y) \geq 0\},$$
$$\Gamma_- = \{y \in \Gamma : h_x(y) < 0\},$$

and construct two trajectories of an isotropic WB-process, $\{y_i^1\}$ and $\{y_i^2\}$, with starting points y_0^1 distributed on Γ_+ with density p_0^1 and y_0^2 distributed on Γ_- with density p_0^2, respectively. These densities must be consistent with the function $h_x(y)$, which, as we know, has an apparent geometric interpretation. Namely,

$$\frac{\sigma_m}{2}|h_x(y)|d\sigma(y) = \frac{|\cos\varphi_{yx}|}{|y-x|^{m-1}}d\sigma(y)$$

is a solid angle of view of a surface element $d\sigma(y)$ from the point x. So we can proceed as follows. Choose an isotropically distributed direction of a ray coming out of the point $x \in G_1$. This ray intersects Γ with probability Ω/σ_m, where

$$\Omega(x) = \int_{\Gamma_+} \frac{\cos\varphi_{yx}}{|y-x|^{m-1}q(x,y)}d\sigma(y) \tag{3.37}$$

is a solid angle of view of the whole surface Γ from the point x.

With a conditional probability 1, there will be even number of intersections. Denote this number by $2q(x,y)$. There will be q points on Γ_+ and q points on Γ_-. In a convex case, $q = 1$ and so we can take these intersections as the initial points y_0^1 and y_0^2, respectively. If a boundary is not convex, there can arise such situations that $q > 1$ and in that case we choose randomly with probability q^{-1} one of the points on Γ_+ and one on Γ_- as y_0^1 and y_0^2, respectively. As a result, we come to the following initial density:

$$p_0^{1(2)}(x;y) = \frac{|\cos\varphi_{yx}|}{\Omega q(x,y)|x-y|^{m-1}}. \tag{3.38}$$

Having sampled initial points, we construct WB-processes $\{y_i^1\}$ and $\{y_i^2\}$ for $i \geq 1$ using the transitional density (3.16). Since this density is consistent with $k(y_{i+1}, y_i)$, an adjoint estimator is natural to be used here. Therefore,

$$\xi^*(x;n) = Q_0^*(x)\sum_{i=0}^{n}(T_i^1 f(y_i^1) - T_i^2 f(y_i^2)) \tag{3.39}$$

is the desired δ-biased estimator for $W(x)$. Coefficients T_i^k are calculated according to formula (3.19) using trajectories of the corresponding Markov chains $\{y_i^k\}, k = 1, 2,$

$$Q_0^*(x) = \frac{h_x(y_0^1)}{p_0^1(y_0^1)} = 2q(x, y_0^1)\frac{\Omega(x)}{\sigma_m}.$$

Two quantities are not defined yet, Ω and λ. Note that ξ^* depends on Ω linearly and so we can use an unbiased estimator instead of its exact value. The geometric algorithm of sampling initial points gives us a clue to construction of this estimator. Here, we have the probability of success (i.e., of intersection) equal to Ω/σ_m if the ray direction is sampled in the whole solid angle. Recall now that all the estimators constructed in this book are used in a common statistical way. It means that a large enough number, N, of independent realizations of ξ^* is constructed, and then they are averaged in order to obtain the desired solution. Denote by N_1 a number of independently sampled random directions, $N_1 > N$. Then N/N_1 is an unbiased estimator for Ω/σ_m. So, instead of $Q_0^\star$, we can use

$$\bar{Q}_0^\star = 2q(x, y_0^1) \cdot \frac{N}{N_1}.$$

To compute the value of α, two limit theorems can be used. It is known that $\lambda_1 = 1$ is the maximum eigenvalue of the integral operator $-K^*$ and μ_0 is the unique eigenfunction corresponding to this value. So we have [32]

$$\mu_0(y) = \lim_{n\to\infty} (-K^*)^n z(y) \cdot \mathrm{const}$$

uniformly over $y \in \Gamma$ for an arbitrary non-zero continuous function z. Using direct estimators constructed for an infinite isotropic WB-process $\{y_i; i = 0, 1, \dots\}$, we come to the following:

$$\alpha = \frac{\lim_{n\to\infty} \mathbb{E}(Q_0 T_n g(y_n))}{\lim_{n\to\infty} \mathbb{E}(Q_0 T_n |y_n|^{2-m})},$$

where

$$Q_0 = \frac{z(y_0)}{p_0(y_0)}.$$

Another way of finding μ_0 and integral functionals of this function follows from the ergodic theorem (Section 6.1). With probability 1, we have

$$\alpha = \frac{\lim_{n\to\infty} \frac{1}{n} \sum_{i=1}^{n} g(y_i)}{\lim_{n\to\infty} \frac{1}{n} \sum_{i=1}^{n} |y_i|^{2-m}}.$$

From the computational point of view, it would be more convenient to rearrange (3.39) in such a way that the parameter α is used in the final expression only. Hence,

$$\xi^\star(x; n) = \alpha \left(\bar{Q}_0^\star(x) \sum_{i=0}^{n} (T_i^1 |y_i^1|^{2-m} - T_i^2 |y_i^2|^{2-m}) \right.$$
$$\left. - \bar{Q}_0^\star(x) \sum_{i=0}^{n} (T_i^1 g(y_i^1) - T_i^2 g(y_i^2)) \right). \tag{3.40}$$

It is essential to note that this estimator has finite variance even in the case when the point x approaches the boundary, Γ, since singularity of the function $h_x(y)$ is included in the distribution density of the initial point. As a consequence, in the neighbourhood of the boundary, a solution of the problem can be evaluated at one point only.

Suppose now that points $x_0, x_1, \dots, x_r$ are separated from the boundary. Universal initial densities p_0^1, p_0^2 from (3.38) cannot be used in this case since fraction

$h_{x_j}(y_0^k)(p_0^k(x_0;y_0^k))^{-1}$ is not uniformly bounded on Γ. Hence, the choice of special densities p_0^k, $k=1,2$ consistent with all $h_{x_j}(y_0^k)$ depends on the geometry of Γ.

3.5 Third BVP and alternative methods of solving the Dirichlet problem

We consider here the third classical BVP for the Laplace equation

$$\begin{aligned} \Delta u(x) &= 0, \quad x \in G \ (\text{or } x \in G_1), \\ \frac{\partial u}{\partial n}(y) - H(y)u(y) &= g(y), \quad y \in \Gamma, \end{aligned} \tag{3.41}$$

where $H \not\equiv 0$ is some continuous function.

Theorem 3.7. *The third BVP is solvable for all continuous functions g, and its solution can be represented in the form of a single-layer potential, if $H \geq 0$ for the interior problem and $H \leq 0$ for the exterior one.*

So, let us suppose that a solution $u(x)$ of the problem (3.41) equals $V[\mu](x)$, a single-layer potential with an unknown density $\mu(y)$. Theorem 3.1 gives us the integral equation of the second kind for this density:

$$\mu(y) = \lambda \int_\Gamma k_H^\star(y,y')\mu(y')d\sigma(y') + f(y), \quad y \in \Gamma, \tag{3.42}$$

where

$$k_H^\star(y,y') = \frac{2}{\sigma_m}\left(\frac{\cos\varphi_{yy'}}{|y-y'|^{m-1}} - \frac{H(y)}{(m-2)|y-y'|^{m-2}}\right), \tag{3.43}$$

$\lambda = 1, f = -g$, for the interior problem,
$\lambda = -1, f = g$, for the exterior one.
On Lyapunov surfaces, the integral operator $K_H^\star$ with the kernel $k_H^\star$ is weakly singular, and it is compact in $C(\Gamma)$. Using the same technique of the Green's formulas and general spectral properties of a compact operator, we come to the following.

Proposition 3.3. *Characteristic values λ_i of (3.42) are real numbers. There exist such positive a and b that, first, if $H \geq 0$, then λ_i are either less than or equal to $-a$, or greater than or equal to $1+b$. Second, if $H \leq 0$, then λ_i are less than or equal to $-(1+b)$, or greater than or equal to a.*

It is worth noting that an actual arrangement of λ_i strongly depends on a boundary configuration and on the function $H(y)$. For example, if Γ is a sphere and $H > 0$ is constant on Γ, then [111]

$$\lambda_i = \frac{2i-1}{1-2H}, \quad i = 1, 2, \ldots.$$

So we see that for $H < 0.5$, all λ_i are positive, $\lambda_i \geq \frac{1}{1-2H} = 1 + b$ and the Neumann series converges. This convergence holds for $H > 1$ also. In this case, $\lambda_i \leq \frac{1}{1-2H} = -a < -1$. If $\frac{1}{2} < H < 1$, then $|\lambda_1| \leq 1$ and those analytical continuation methods that have been used in Chapter 3.2 for the Dirichlet BVP are applicable here.

In a general case, it is known [87] that for sufficiently small H, $|\lambda_1|$ is greater than 1, and hence the Neumann series converges. What's more, this convergence holds even if we substitute the kernel k_H^* by its modulus. So we can use the standard method of a Monte Carlo estimator construction with a non-zero probability of a Markov chain termination. However, in every concrete case, we have to calculate the value of a beforehand, while it is not necessary to know the exact value of λ_1. Suppose that $1 \geq |\lambda_1| \geq a$, where a can be calculated by using some iterative procedure, a Monte Carlo algorithm, in particular [23]. The spectral parameter substitution $\lambda = \lambda_0 \frac{2a\eta}{1-\eta}$, where $\lambda_0 = 1$ for the interior problem and $\lambda_0 = -1$ for the exterior one, and corresponding rearrangement of the integral equation lead to the representation of a solution in the form of a convergent series for the integral operator

$$K_1^* = \eta^* I + 2a\eta^* \lambda_0 K_H^*,$$

where $\eta^* = \frac{1}{1+2a}$ (Chapter 2.2). Both scalar and vector algorithms with a resolvent iteration can be used here also, and a can be taken as a parameter of these methods. The substitution (Chapter 2.3)

$$\lambda = \frac{4a\eta}{(1-\eta)^2}$$

also can be used here, since images of points lying on the ray $\lambda \geq 1 + b$ have moduli, which are greater than

$$\eta^* = 1 + 2a - 2\sqrt{a^2 + a}.$$

Suppose now that we know both limits of the characteristic set for the operator K_H^*. Then the following spectral parameter substitutions can be used here:

$$\lambda = \frac{4a(1+b)\eta}{(1+b)(1-\eta)^2 + a(1+\eta)^2}$$

for the interior problem, and

$$\lambda = \frac{4a(1+b)\eta}{a(1-\eta)^2 + (1+b)(1+\eta)^2}$$

for the exterior problem. These functions conformally map the unit disk, $\{\eta : |\eta| < 1\}$, onto the complex plane with two rays excluded: $\lambda \leq -a$, $\lambda \geq 1 + b$ and $\lambda \leq -1 - b$, $\lambda \geq a$, respectively.

Finally, all the considered transformations of the initial Neumann series lead to a representation of a solution in the form of a finite sum of integral functionals:

$$u(x) = \left(\sum_{i=0}^{n} l_i^{(n)} \lambda^i K_H^{*i} f, h_x^* \right) + \delta(x;n)$$
$$= \sum_{i=0}^{n} l_i^{(n)} \left(f, \lambda^i K_H^i h_x^* \right) + \delta(x;n).$$

To construct a Monte Carlo estimator based on this representation, we have to choose a Markov chain with a transitional density consistent with the kernel k_H^*, or what is equivalent consistent with the kernel k_H. To start with, consider the density of the isotropic WB-process. Let Γ be convex, then

$$\frac{k_H^*(y,y')}{p(y',y)} = 1 - \frac{H(y)|y-y'|}{(m-2)\cos\varphi_{yy'}}, \tag{3.44}$$

and this weight coefficient can be unbounded. For example, this can happen in the case when Γ has plane parts. Hence, we come to a conclusion that the isotropic WB-process is applicable for solving the third BVP only in the case of strictly convex boundaries.

Consider a direct estimator now:

$$\xi_H(x;n) = \sum_{i=0}^{n} l_i^{(n)} \lambda^i Q_{H,i} h_x^*(y_i), \tag{3.45}$$

where

$$Q_{H,0} = \frac{f(y_0)}{p_0(y_0)}, \quad Q_{H,i} = Q_{H,i-1} \frac{k_H^*(y_i, y_{i-1})}{p(y_{i-1}, y_i)},$$
$$h_x^*(y_i) = \frac{2}{(m-2)\sigma_m} \cdot \frac{1}{|x-y_i|^{m-2}}$$

and $\{y_0, y_1, y_2, \ldots\}$ is a Markov chain of WB-process constructed in accordance with some initial p_0 and transitional p densities. We see that conditions for p_0 are not very strict and it is sufficient to have $\alpha_2 \geq p_0(y_0) \geq \alpha_1 > 0$ for some constant α_1, α_2. As for the transitional density, we can proceed as follows.

Let $y_{i-1} \in \Gamma$ be already defined. Since $k_H^*(y_i, y_{i-1})$ is weakly singular as y_i approaches y_{i-1}, then to have $Q_{H,i}$ bounded, we must include this singularity into the transitional density. So we divide a boundary surface into two parts: some neighbourhood C of the point y_{i-1}, $C = \{y : |y - y_{i-1}| < r_0\}$ and all the remaining boundary $\Gamma \setminus C$. With some probability ω, a point y_i is sampled in C in accordance with some singular density p, and with the complementary probability $1-\omega$, we use some arbitrary density $\pi(y) \geq \alpha > 0$ to sample the next point in $\Gamma \setminus C$. The choice of a singular density in C depends on the geometric properties of a surface Γ at the point y_{i-1}.

Let $m = 3$ for simplicity. With probability 1, y_{i-1} is either an elliptic point or an interior point of a flat part of the boundary. Suppose that curvature of Γ at this point is finite. From here, we can conclude that the fraction in (3.44) is a bounded function of y in C for some r_0, $y' = y_{i-1}$. It means that the isotropic transitional density can be used in C in this case, and the resulting transitional density on the whole Γ is

$$p_1(y_{i-1}, y_i) = \omega p(y_{i-1}, y_i) \cdot I_C(y_i) + (1-\omega)\pi(y_i) I_{\Gamma \setminus C}(y_i), \tag{3.46}$$

where

$$I_C(y) = \begin{cases} 1, & y \in C \\ 0, & y \notin C \end{cases}$$

is the indicator function of the set C.

Let y_{i-1} lie on a flat part of Γ. Introduce the local polar coordinate system in this plane setting $y_{i-1} = (0, 0)$ and $y_i = (r, \varphi)$. Obviously, there exists such r_0 that C is a circle lying in this plane. We can choose the function $(2\pi r_0 |y_i - y_{i-1}|)^{-1}$ as a density in this circle, and so we have

$$p_1(y_{i-1}, y_i) = \omega (2\pi r_0 |y_i - y_{i-1}|)^{-1} \cdot I_C(y_i) + (1-\omega)\pi(y_i) I_{\Gamma \setminus C}(y_i). \tag{3.47}$$

It means that with probability ω, $y_i \in C$, and the weight coefficient is equal to

$$\frac{k_H^*(y_i, y_{i-1})}{p_1(y_{i-1}, y_i)} = -\frac{1}{\omega} H(y_i) r_0,$$

since $\cos \varphi_{y_i y_{i-1}} = 0$ here, and with probability $1-\omega$, $y_i \in \Gamma \setminus C$, and

$$\frac{k_H^*(y_i, y_{i-1})}{p_1(y_{i-1}, y_i)} = \frac{1}{1-\omega} \cdot \frac{k_H^*(y_i, y_{i-1})}{\pi(y_i)}.$$

In Chapter 3.4, we considered the exterior Dirichlet problem and represented its solution as a sum (3.34) of two functions: a double-layer potential with a density μ and the Robin's potential. The last one is equal to $O(|x|^{2-m})$ when $|x| \to \infty$. This feature ensures the most general harmonic properties at infinity for the solution of BVP (3.31). The same result can be obtained if instead of the Robin's potential we use a single-layer potential with the density equal to $-H\mu$, where H is some continuous function. In particular, H may be equal to some non-zero constant value on Γ. So, consider the Dirichlet problem and represent its solution in the form

$$u(x) = W[\mu](x) + V[-H\mu](x). \tag{3.48}$$

As a consequence, we come to the integral equation

$$\mu = \lambda K_H \mu + f,$$

where K_H is the integral operator adjoint to K_H^*. Its kernel $k_H(y, y')$ is equal to $k_H^*(y', y)$, which is defined in (3.43). For the interior problem, $\lambda = -1$ and $f = g$, while for the exterior one, $\lambda = 1$ and $f = -g$.

From Statement 3.2, it follows that the spectral properties of the integral operator K_H are identical to the spectral properties of K_H^*. Hence, the following proposition is valid.

Theorem 3.8. *The Dirichlet problem is solvable for all continuous boundary values g and its solution can be represented as a sum of two potentials, $W[\mu]$ and $V[-H\mu]$, where $H \le 0$ for the interior problem and $H \ge 0$ for the exterior one.*

Since $k_H(y, y') = k_H^*(y', y)$, it means that the same WB-process that has been constructed for solving the third BVP is applicable here. The main difference is that we use the adjoint estimator here:

$$\xi_H^*(x; n) = \sum_{i=0}^{n} l_i^{(n)} \lambda^i Q_{H,i}^* f(y_i), \tag{3.49}$$

where

$$Q_{H,0}^* = \frac{1}{p_0(y_0)} \cdot \frac{2}{\sigma_m} \left(\frac{\cos \varphi_{y_0 x}}{|y_0 - x|^{m-1}} - \frac{H(y_0)}{(m-2)|y_0 - x|^{m-2}} \right),$$

$$Q_{H,i}^* = Q_{H,i-1}^* \frac{k_H(y_{i-1}, y_i)}{p(y_{i-1}, y_i)},$$

and the transitional density, p, is either isotropic for strictly convex boundary or it is equal to p_1 defined in (3.46), (3.47).

Another alternative method of solving Dirichlet BVPs is based on the representation of a solution in the form of a single-layer potential. As a result, we come to the need of solving boundary integral equation of the first kind:

$$\int_{\Gamma} \frac{1}{(m-2)\sigma_m} \cdot \frac{\mu(y')}{|y - y'|^{m-2}} \, d\sigma(y') = g(y), \quad y \in \Gamma. \tag{3.50}$$

To construct a stochastic estimator for a solution of the Dirichlet problem (for exterior and interior simultaneously), we can use the results of Section 2.5. Here, we have

$$u(x) = (h_x, \mu),$$

where

$$h_x(y) = \frac{1}{(m-2)\sigma_m} |x - y|^{2-m},$$

and

$$k(y,y') = k(y',y) = \frac{1}{(m-2)\sigma_m} |y-y'|^{2-m}.$$

Note that the transitional densities (3.46), (3.47) are consistent with this kernel. Hence, they can be used for construction of the estimator.

3.6 Inhomogeneous problems

Up to this point, everywhere in Chapter 3, the Laplace equation has been considered. Suppose now that its right-hand side is non-zero, and so we come to the need of solving some BVP for the Poisson's equation:

$$\begin{aligned} \Delta u(x) &= -F(x), \quad x \in G, \\ Bu(y) &= g(y), \quad y \in \Gamma, \end{aligned} \tag{3.51}$$

where B is a boundary operator of the first ($B = I$), second ($B = \frac{\partial}{\partial n(y)}$), or third BVP ($B = \frac{\partial}{\partial n(y)} - H(y)$).

Consider Newton's volume potential,

$$V_0(x) = \int_G \frac{1}{(m-2)\sigma_m} \cdot \frac{F(y)}{|x-y|^{m-2}} dy. \tag{3.52}$$

Let $F(y)$ be continuous function in a bounded domain G. Then $V_0 \in C^1(\mathbb{R}^m)$, it is harmonic in G_1, and $V_0(x) = O(|x|^{2-m})$ for $|x| \to \infty$. It is well known [120] that even in the case when F is finitary distribution (generalized function), the potential V_0 satisfies Poisson's equation

$$\Delta V_0 = -F,$$

and if $F \in C^1(G) \cap C(\bar{G})$, then this equation is satisfied at every point $x \in G$. As a result, we can proceed in the following way. Consider BVP (3.51) for u and represent its solution as a sum of two functions:

$$u(x) = v(x) + V_0(x).$$

Hence, the auxiliary function $v(x)$ satisfies the Laplace equation, and it is a solution of the correspondent BVP

$$\begin{aligned} \Delta v(x) &= 0, \quad x \in G, \\ Bv(y) &= g(y) - BV_0(y), \quad y \in \Gamma. \end{aligned} \tag{3.53}$$

It is essential to note that we have to know boundary values of the integral (3.52) or its normal derivative at every point y on the boundary. In contrast to deterministic approach, when considering BVP (3.51) in the context of Monte Carlo methods, we can easily overcome this difficulty. The double randomization principle makes it possible to use a random estimate instead of exact value of a multidimensional integral BV_0.

These considerations are valid for the exterior BVP as well,

$$\Delta u(x) = -F(x), \quad x \in G_1,$$
$$Bu(y) = g(y), \quad y \in \Gamma.$$

Some additional restrictions must be imposed on the function F here. We can require, for example, that it has a compact support.

Consider another approach to solving BVP (3.51). Suppose that we construct such sufficiently smooth function $u_2(x)$ that at every point $y \in \Gamma$,

$$u_2(y) = g(y).$$

To find such function, one of the possible ways could be the $R-$ function method (for details, see [98]).

Next we consider Dirichlet BVP (3.51) for Poisson's equation and represent its solution in the form of a sum

$$u(x) = u_1(x) + u_2(x), \quad x \in G. \tag{3.54}$$

As a result, we come to the BVP for the Poisson's equation with zero boundary values

$$\Delta u_1(x) = -F(x) - \Delta u_2(x), \quad x \in G,$$
$$u_1(y) = 0, \quad y \in \Gamma.$$

Another way of transition from BVP (3.51) to a homogeneous BVP for the Poisson's equation is following. We can consider function $u_2(x)$ to be a solution to the correspondent BVP for the Laplace equation

$$\Delta u_2(x) = 0, \quad x \in G,$$
$$Bu_2(y) = g(y), \quad y \in \Gamma. \tag{3.55}$$

Then, we have

$$\Delta u_1(x) = -F(x), \quad x \in G,$$
$$Bu_1(y) = 0, \quad y \in \Gamma.$$

To solve this problem, the following algorithm can be proposed. Let $u_\delta(x, x_0)$ be the influence function, i.e., for every $x_0 \in G$,

$$\Delta_x u_\delta(x, x_0) = \delta(x - x_0), \quad x \in G, \quad x \neq x_0,$$
$$B_y u_\delta(y, x_0) = 0, \quad y \in \Gamma.$$

We see that u_δ is the Green's function in the case of the Dirichlet BVP. It is obvious that function u_1 can be represented in the form of an integral functional

$$u_1(x) = -(u_\delta, F) = -\int_G u_\delta(x, x_0)F(x_0)dx_0. \tag{3.56}$$

Represent u_δ as a sum of the fundamental solution for the Laplace equation and some function u_0:

$$u_\delta(x, x_0) = -\frac{1}{(m-2)\sigma_m}|x - x_0|^{2-m} + u_0(x, x_0). \tag{3.57}$$

Then, we have that u_0 is a solution to the BVP for the Laplace equation

$$\begin{aligned} &\Delta_x u_0(x, x_0) = 0, \quad x \in G, \\ &B_y u_0(y, x_0) = \frac{1}{(m-2)\sigma_m} B_y(|y - x_0|^{2-m}), \quad y \in \Gamma. \end{aligned} \tag{3.58}$$

So we see that a solution of the problem (3.51) is represented as a linear functional of solutions u_2 and u_0 to the BVPs for the Laplace equations (3.55) and (3.58), respectively. Note that the inverse transition from u_2 and u_0 does not present any challenge to Monte Carlo methods. Denote by $\xi_0(x, x_0)$ a random estimator for $u_0(x, x_0)$, and by $\xi_2(x)$, we denote a random estimator for $u_2(x)$. Randomizing the integration in (3.56), we choose a random point x_0 in G with a probability density $\pi(x_0)$. Then, according to the double randomization principle, we have that

$$\xi(x) = \xi_2(x) + \frac{F(x_0)}{\pi(x_0)}\left(\frac{1}{(m-2)\sigma_m}|x - x_0|^{2-m} - \xi_0(x, x_0)\right)$$

is an estimator for a solution of BVP (3.51) for Poisson's equation. Properties of this estimate (unbiasedness, consistency, finiteness of variance, etc.) are obviously determined by the correspondent properties of the random estimates ξ_0, ξ_2 for a solution of the Laplace equation.

3.7 Continuity BVP

Consider the problem of computing a function, u, which satisfies the Poisson equation inside a compact domain, G:

$$\epsilon_i \Delta u(x) = -\rho(x), \ x \in G, \tag{3.59}$$

the Laplace equation outside the domain:

$$\epsilon_e \Delta u(x) = 0, \ x \in \mathbb{R}^m \setminus \overline{G}, \tag{3.60}$$

and the continuity conditions on the boundary of G:

$$u_i(y) = u_e(y), \ \epsilon_i \frac{\partial u_i}{\partial n(y)} = \epsilon_e \frac{\partial u_e}{\partial n(y)}, \ y \in \partial G. \tag{3.61}$$

Here, u_i and u_e are the limit values of the solution from inside and outside, respectively; ϵ_i and ϵ_e are constants such that $\epsilon_i < \epsilon_e$.

With the assumption that ∂G is smooth enough, it is possible to represent the solution as a sum of a volume and single-layer potentials [32, 34],

$$u(x) = g(x) + \int\limits_{\partial G} \frac{2}{\sigma_m} \frac{1}{|x-y|^{m-2}} \mu(y) d\sigma(y) \equiv g(x) + u_0(x), \tag{3.62}$$

where $g(x) = \int_G \frac{1}{(m-2)\sigma_m} \frac{\rho}{\epsilon_i |x-y|^{m-2}} \, dy$.

Taking into account boundary conditions (3.61) and discontinuity properties of the single-layer potential's normal derivative [34], we arrive at the integral equation for the unknown density, μ:

$$\mu = -q_0 \mathcal{K} \mu + f, \tag{3.63}$$

which is valid almost everywhere on ∂G. Here, $q_0 = \frac{\epsilon_e - \epsilon_i}{\epsilon_e + \epsilon_i}$, and the kernel of the integral operator $\mathcal{K}$ is $\frac{2}{\sigma_m} \frac{\cos \phi_{yy'}}{|y-y'|^{m-1}}$, where $\phi_{yy'}$ is the angle between (NOTE!) **external** normal vector $n(y)$ and $y - y'$. The free term of this equation equals

$$f = q_0 \frac{\partial g}{\partial n(y)},$$

and it can be computed analytically. Since $q_0 < 1$, the Neumann series for (3.63) converges (see, e.g. [32, 34]), and it is possible to calculate the solution as

$$u_0(x) = \sum_{i=0}^{\infty} (h_x, (-q_0 \mathcal{K})^i f), \tag{3.64}$$

where $h_x(y) = \frac{2}{\sigma_m} \frac{1}{|x-y|^{m-2}}$. Usually, however, $\epsilon_e \gg \epsilon_i$ and, hence, $||\lambda_1| - 1| = \frac{2\epsilon_i}{\epsilon_e + \epsilon_i} \ll 1$. Here, $\lambda_1 = -1/q_0 > 1$ is the smallest characteristic value of the operator $-q_0 \mathcal{K}$. This means that convergence in (3.64) is rather slow, even for convex G.

To speed up the convergence in (3.64), we apply the method of spectral parameter substitution considered in the previous sections (see, e.g. [46, 98, 104] for Monte Carlo algorithms based on this method). This means we consider the parameterized equation

$$\mu_\lambda = \lambda(-q_0 \mathcal{K}) \mu_\lambda + f$$

and analytically continue its solution given by the Neumann series for $|\lambda| < |\lambda_1|$. This goal can be achieved by substituting in λ its analytical expression in terms of another complex parameter, η, and representing μ_λ as series in powers of η.

In this particular case, it is possible to use the substitution $\lambda = \frac{2|\lambda_1|\eta}{1-\eta} \equiv \chi(\eta)$, and hence

$$u_0(x) = \sum_{i=0}^{n} l_i^{(n)}(-q_0)^i \left(h_x, \mathcal{K}^i f\right) + O(\gamma^{n+1}), \tag{3.65}$$

where $\gamma = \frac{1}{1+2|\lambda_1|} < \frac{1}{3}$ and $l_i^{(n)} = \sum_{j=i}^{n} C_{j-1}^{i-1}(2|\lambda_1|)^i \gamma^j$. The rate, γ, of the geometric convergence of the transformed series in powers of η at the point $\eta_0 = \chi^{-1}(1)$ is determined by the ratio of $|\eta_0|$ and $L = \min_i |\chi^{-1}(\lambda_i)|$. Here, λ_i are characteristic values of $\mathcal{K}$ (and $\mathcal{K}^*$) and $L = 1$ [34, 46].

Given a desired computational accuracy, we can calculate the number of terms needed in (3.65). Thus, the problem reduces to computing a finite number of multidimensional integrals.

3.7.1 Walk on boundary for the continuity problem

To construct Monte Carlo estimates for $u_0(x)$, it is sufficient to calculate the integral functionals $I_i(x) = \left(h_x, \mathcal{K}^i f\right)$ of iterations of the integral operator. Here, the domain of integration is $[\partial G]^{i+1}$.

Let G be a convex domain. In this case, the kernel, $k(y, y')$, of $\mathcal{K}$ corresponds to the uniform in a solid angle distribution of the point y as viewed from the point y'. This means that the most convenient way to construct a random estimate for double integral $(h_x, \mathcal{K}f)$ is to use for choosing y' some density p_0 that conforms with f and to sample y in accordance with the conditional density $p(y' \to y) = k(y, y')$. This construction leads to using so-called 'direct' estimates for $I_i(x)$ [104]. Therefore,

$$I_i(x) = \mathbb{E} Q_i h_x(y_i), \tag{3.66}$$

where $Y = \{y_0, y_1, \dots\}$ is the Markov chain of random points on the boundary ∂G, with the initial density p_0 and the transition density $p(y_i \to y_{i+1}) = \frac{2}{\sigma_m} \frac{\cos \phi_{y_{i+1} y_i}}{|y_{i+1} - y_i|^{m-1}}$. Here, random weights are $Q_i = Q_0 = \frac{f(y_0)}{p_0(y_0)}$, $i = 1, 2, \dots, n$. Therefore, $O(\gamma^{n+1})$-biased estimate for u_0 is

$$\xi = \sum_{i=0}^{n} l_i^{(n)}(-q_0)^i Q_i h_x(y_i). \tag{3.67}$$

Consider now a non-convex G. In this case, first, there is a possibility of multiple intersections of a straight line with the boundary, and, second, $\cos \phi_{y_{i+1} y_i}$ can be negative. There are two different approaches to overcome this difficulty. Let $q_i(y_i)$ be the number of these crossings, y_i excluded. The first solution is to choose the next point of Markov chain y_{i+1} randomly from q_i intersections. This corresponds to the transition density $p_1 = \frac{1}{q_i}|p|$ and therefore $Q_{i+1} = Q_i\ q_i\ \text{sign}(\cos\phi_{y_{i+1}y_i})$. The second approach is to use a branching Markov chain, thus taking into account all intersections. In this case, the transition density is $p_2 = |p|$, weights $Q_{i+1} = Q_i\ \text{sign}(\cos\phi_{y_{i+1}y_i})$ can change only their signs but not their moduli, and every point y_i gives birth to $q_i(y_i)$ points in the next generation.

There is no exact criterion that could determine which of these two approaches is more efficient. On the one hand, the random choice of the next point leads to a fixed length of the Markov chain, which results in smaller computational time needed to construct a sample of the solution's estimate. On the other hand, the weight factor in this approach can be as large as $\prod_1^{n+1} q_i$, and this could result in large variance. The estimate based on the construction of a branching Markov chain has the opposite properties. Branching could result in small variance, but it also requires longer computations needed to obtain an estimate's sample.

In both approaches, the construction of the Markov chain, Y, is based on its geometrical interpretation. Given a point, y_i, we simulate a random isotropic direction ω_i and find the next point y_{i+1} as the intersection of this direction with the boundary surface ∂G. It is well known that different procedures can be used to choose ω_i, and a special caution is needed if quasirandom numbers are planned to be used (see, e.g. [40]).

3.8 Calculation of the solution derivatives near the boundary

Consider a general BVP for the Laplace equation

$$\begin{aligned} \Delta u(x) &= 0, \quad x \in G \quad (\text{or } G_1), \\ Bu(y) &= g(y), \quad y \in \Gamma \end{aligned} \tag{3.68}$$

and suppose that we seek for its solution in the form of a single-layer potential. Here, B is one of the classical boundary operators. Depending on the concrete form of this operator, we come to the boundary integral equation of the second or the first kind. As we have already shown in this chapter, it provides a possibility to construct some biased estimator with a help of WB-process.

Here, we will consider three-dimensional (3D) BVP, but the same reasoning is valid in the general m-dimensional case, $m = 2$ in particular. So we have

$$u(x) = \int_\Gamma \frac{1}{2\pi} \cdot \frac{\mu(y)}{|x-y|} d\sigma(y),$$

and, as is already shown in the previous sections, we can write out the general representation for the unknown density of the potential

$$\mu(y) = \sum_{i=0}^{n} a_i^{(n)} K^i f(y) + \varepsilon(y; n), \tag{3.69}$$

where K is an integral operator and f is a free term of the corresponding integral equation. As for coefficients $a_i^{(n)}$ and a bias function ε, they are determined by the concrete BVP and a chosen method of the integral solution's representation (Sections 3.2–3.5).

To calculate the derivatives of $u(x)$ near the boundary, some estimator for the value of density $\mu(y)$ at the point y on Γ will be needed. As we know (Chapter 2), an adjoint Monte Carlo estimator is natural to be used here. Let $\{z_0 = y, z_1, z_2, \dots\}$ be WB-process with a transitional density $p(z_{i-1}, z_i)$ consistent with the integral operator kernel $k(z_{i-1}, z_i)$. Then, the ε-biased estimator for $\mu(y)$ is

$$\xi^*(y) = \sum_{i=0}^{n} a_i^{(n)} Q_i^* f(z_i), \tag{3.70}$$

where

$$Q_0^* = 1, \quad Q_i^* = Q_{i-1}^* \frac{k(z_{i-1}, z_i)}{p(z_{i-1}, z_i)}.$$

It has been already mentioned that the isotropic transitional density is not always consistent with a kernel $k(z_{i-1}, z_i)$, but methods of attaining this consistency are also described in the previous sections.

For definiteness, let $x = (x_1, x_2, x_3)$ lie in the interior of G. Then

$$\frac{\partial u}{\partial x_i}(x) = \int_\Gamma \frac{1}{2\pi} \cdot \frac{y_i - x_i}{|x-y|^3} \mu(y) d\sigma(y), \quad i = 1, 2, 3.$$

Denote by $t \in \Gamma$ the nearest to x point on the boundary and suppose that t is either an elliptic point or the inner point of the flat part of the boundary. Hence, there exists the definite tangent plane in this point and the local orthonormal coordinate system based on the normal vector $n(t)$ and the pair of tangent vectors τ_1, τ_2 can be introduced. So we have

$$\begin{aligned} \frac{\partial u}{\partial x_i}(x) &= \frac{\partial u(x)}{\partial n(t)} n_i(t) + \sum_{j=1}^{2} \frac{\partial u(x)}{\partial \tau_j(t)} \tau_{ji}(t) \\ &= \frac{\partial u}{\partial n} n_i + \frac{\partial u}{\partial \tau}. \end{aligned} \tag{3.71}$$

Here, $\tau = \tau_1 \tau_{1i} + \tau_2 \tau_{2i}$. Representation (3.71) makes it possible to calculate separately $\partial u / \partial \tau$, which becomes singular integral in the limit when x tends to the boundary. We proceed as follows (Chapter 5).

Sample a random point $y \in \Gamma$ according to some density $p(y)$. Introduce local spherical coordinates with the origin at the point t in such a way that $y = (r, \theta, \psi)$, where $r = |y - t|$, θ is the angle between the vector $y - t$ and the tangent at the point t plane and ψ is the angle between the projection $\pi(y - t)$ of this vector on the tangent plane and τ_1. Define now symmetrical with respect to t points lying on the boundary Γ:

$$\chi_t(y) = (\bar{r}, \theta, \psi + \pi),$$

and

$$\alpha_t(y) = (r, \bar{\theta}, \psi + \pi),$$

and set $y' = \chi_t(y)$ if t is an elliptic point and $y' = \alpha_t(y)$ if t lies on the flat part of the boundary. Here, $\bar{r}$ and $\bar{\theta}$ are determined by the condition that y' lies on Γ. Then, we have

$$\frac{\partial u(x)}{\partial \tau(t)} = \mathbb{E}\zeta_\tau(x) + \varepsilon^{(\tau)}(x), \quad |\varepsilon^{(\tau)}| \le \text{const} \cdot \varepsilon_0,$$

where

$$\zeta_\tau(x) = \frac{1}{4\pi}\left(\frac{\cos(\tau, y - x)}{|y - x|^2}\xi^*(y) + \frac{\cos(\tau, y' - x)}{|y' - x|^2}\xi^*(y')\right) \cdot \frac{1}{p(y)}, \tag{3.72}$$

and $\varepsilon_0 = \sup_{y \in \Gamma} |\varepsilon(y; n)|$, $\cos(\tau, y - x)$ is cosine of angle between τ and $y - x$.

In order to prove that the variance of ζ_τ is finite, we consider this estimator for $|y - t| \to 0$. We have

$$\zeta_\tau(x) = \frac{1}{4\pi} \cdot \frac{\cos(\tau, y - x)}{|y - x|^2}\left(\xi^*(y) - \xi^*(y') + \xi^*(y')\left(1 - \frac{|\pi(y - x)|}{|\pi(y' - x)|}\right)\right)\frac{1}{p(y)}.$$

It is obvious that

$$|\pi(y' - x)| - |\pi(y - x)| = |\pi(y' - x)| \cdot O(|y - t|),$$

if t is an elliptic point and equal zero if t, y, y' lie on the same plain. Suppose that f is Hölder continuous with an exponent β on Γ. Then, we can construct estimators $\xi^*(y)$ and $\xi^*(y')$ in such a dependent way that

$$|\xi^*(y) - \xi^*(y)| = O(|t - y|^\beta), \quad |y - t| \to 0.$$

Hence, we have

$$\zeta_\tau(x) = \frac{1}{p(y)} O(|t - y|^{\beta - 2}), \text{ for } |y - t| \to 0.$$

It means that $\mathbb{E}\zeta_\tau^2(x) < \infty$ if density p is singular at the point t with an exponent γ: $p(y) = O(|t - y|^{-\gamma})$, where $\gamma > 2(1 - \beta)$.

Next, consider the normal component of the derivative and denote

$$\frac{\partial u(x)}{\partial n(t)} = \int_\Gamma g(x, t, y)\mu(y)d\sigma(y),$$

where

$$g(x,t,y) = \frac{1}{2\pi}\frac{\cos(n(t), y-x)}{|y-x|^2}.$$

Divide as usual Γ into two parts: Γ_0 and $\Gamma\setminus\Gamma_0$, where

$$\Gamma_0 = \{y : y \in \Gamma, |x-y| = l < l_0\}$$

for some fixed $l_0 > |x-t|$. With some probability α_0, we choose y on Γ_0, and with probability $1-\alpha_0$, we simulate y on $\Gamma\setminus\Gamma_0$.

First, consider t lying in the plain, and suppose that Γ_0 is totally contained in this plain. Denote $\delta = |x-t|$. Then,

$$\int_{\Gamma_0} g(x,t,y)\mu(y)d\sigma(y) = \int_0^{2\pi} d\psi \int_\delta^{l_0} \frac{\delta}{2\pi l^2} g(x,t,l,\psi)\mu(l,\psi)dl.$$

The sampling procedure directly follows from this representation. We choose the angle ψ uniformly in $[0,2\pi]$ and

$$l = \frac{l_0\delta}{l_0(1-\alpha)+\delta\alpha},$$

where α is a random number uniformly distributed in $[0,1]$. It means that the distribution density of the point y is equal to

$$\frac{\delta l_0}{2\pi l^2(l_0-\delta)}.$$

So we have

$$\begin{aligned}\frac{\partial u(x)}{\partial n(t)} &= \mathbb{E}\zeta_n(x) + \varepsilon^{(n)}(x),\\ \zeta_n(x) &= \frac{1}{\alpha_0}\cdot\left(1-\frac{\delta}{l_0}\right)\xi^*(y), \quad y \in \Gamma_0\\ &= \frac{1}{1-\alpha_0}\cdot\frac{g(x,t,y)}{p'(y)}\xi^*(y), \quad y \in \Gamma\setminus\Gamma_0,\end{aligned} \tag{3.73}$$

where $p'(y)$ is a density consistent with g.

Consider the elliptic case now and denote by $R = R(\psi)$ the curvature radius of the surface Γ at the point t in the plain $P(\psi)$. This plain contains $n(t)$, and the angle between $P(\psi)$ and $\tau_1(t)$ is equal to ψ. Consider a surface Γ_1 defined as follows: its section with $P(\psi)$ is a circle of radius $R(\psi)$ with its centre at the point $t + n(t)R(\psi)$.

Then,

$$g(x,t,y)\frac{d\sigma(y)}{dl d\psi} = g(x,t,y_1)\frac{d\sigma_1(y_1)}{dl_1 d\psi}(1+O(l)), \text{ for } l \to 0. \tag{3.74}$$

Here, $y_1 \in \Gamma_1$, $y \in \Gamma_0$ and $l_1 = |x - y_1| = l$, σ_1 is the surface measure on Γ_1. Expression in the right-hand side of (3.74) can be written out explicitly:

$$\frac{l_1^2 + \delta(2R+\delta)}{4\pi R(R+\delta)^2 l_1^2}.$$

So, if we want to obtain an estimator with finite variance, it is necessary to sample the point y on Γ_0 in accordance with the density

$$p_1 = B\frac{l^2 + \delta(2R+\delta)}{2\pi l^2},$$

where B is a normalizing constant. It means that angle ψ is chosen uniformly in $[0, 2\pi]$, next radius $R(\psi)$ is calculated and then l is sampled in accordance with the formula

$$l = -R + \frac{\alpha}{2B} + \left((R - \frac{\alpha}{2B})^2 + \delta(2R+\delta)\right)^{1/2}.$$

Finally, we have

$$\begin{aligned}\zeta_n(x) &= \frac{1}{\alpha_0 p_1}\xi^*(y), \quad y \in \Gamma_0 \\ &= \frac{1}{1-\alpha_0} \cdot \frac{g(x,t,y)}{p'(y)}\xi^*(y), \quad y \in \Gamma \setminus \Gamma_0.\end{aligned} \tag{3.75}$$

Next we substitute (3.72), (3.73), (3.75) in (3.71) and see that the proper choice of the modelling procedures provides a possibility to construct Monte Carlo estimators for the potential derivatives even near the boundary.

3.9 Normal derivative of a double-layer potential

In contrast to other sections, here we consider the 2D case.

Suppose that we seek for solution of a BVP in the form of a double-layer potential with density μ:

$$\begin{aligned}W(x) &= \int_\Gamma \mu(y')\frac{\partial g}{\partial n(y')}(x,y')d\sigma(y') \\ &= \int_\Gamma \left(n_1(y')\frac{\partial g}{\partial y_1'} + n_2(y')\frac{\partial g}{\partial y_2'}\right)(x,y')\mu(y')d\sigma(y').\end{aligned}$$

Here

$$g(x,y) = \frac{1}{\pi} \ln \frac{1}{|x-y|}.$$

Frequently, it is desired to find not only a solution but also its derivatives, and a normal derivative, in particular. To calculate this function, we apply the technique described in [34].

Let $x \in G$ and $y \in \Gamma$; then, we have

$$\frac{\partial W}{\partial x_1} = \int_\Gamma \left(n_1(y') \frac{\partial}{\partial y'_2} - n_2(y') \frac{\partial}{\partial y'_1} \right) \frac{\partial g}{\partial y'_2}(x,y') \mu(y') d\sigma(y'), \tag{3.76}$$

since

$$\frac{\partial g}{\partial x_i} = -\frac{\partial g}{\partial y'_i}$$

and

$$\Delta_x g = 0.$$

Following [34], we introduce differential combinations

$$D_i \mu = B \frac{\partial f}{\partial y'_i} - \frac{\partial f}{\partial n(y')} n_i(y'),$$

where f is a function defined in $\overline{G}$, continuously differentiable in this closed domain, and satisfying boundary conditions $Bf = \mu$. Here, we denote by Bf the boundary value of the function f.

Next, we integrate (3.76) by parts and use the fact that

$$d\sigma(n_1 D_2 - n_2 D_1) f = \left(-dy'_1 \frac{\partial}{\partial y'_1} - dy'_2 \frac{\partial}{\partial y'_2} \right) f = -df$$

is the total differential, and therefore integral of df along a closed curve is equal to zero.

So we have

$$\frac{\partial W}{\partial x_1} = \int_\Gamma \frac{\partial g}{\partial n(y')}(x,y') D_1 \mu d\sigma(y') - \int_\Gamma n_1(y') \left(\frac{\partial g}{\partial y'_1} D_1 \mu + \frac{\partial g}{\partial y'_2} D_2 \mu \right) d\sigma(y'),$$

and as a consequence, after transition to the local coordinate system based on the normal and tangent vectors,

$$\frac{\partial W}{\partial n(y)} = \lim_{x \to y} \int_\Gamma \left(\left(n(y) \cdot \left(\frac{\partial g}{\partial y'_2}(x,y'), -\frac{\partial g}{\partial y'_1}(x,y') \right) \right) \frac{\partial \mu}{\partial \tau(y')} \right) d\sigma(y')$$

since

$$D_i f = \tau_i \frac{\partial f}{\partial \tau}$$

in the 2D case. Here, we use the notations $n(y) = (n_1(y), n_2(y))$ and $\tau(y) = (n_2, -n_1)$.

Finally, if we use the limiting properties of a single-layer potential's derivatives [34],

$$\frac{\partial W}{\partial n(y)} = \int_{\Gamma} \frac{\partial \mu}{\partial \tau(y')} \cdot \frac{\partial g}{\partial \tau(y)}(y, y') d\sigma(y'). \tag{3.77}$$

So, we have represented the normal derivative of a double-layer potential in the form of a singular integral (3.77). Monte Carlo algorithms for calculating such integrals are given in Chapter 5. 3D case is considered there; however, it is essential that the main idea of constructing an estimator based on a pair of symmetrical points is applicable in the 2D case also.

Introduce a local coordinate system at the point y in such a way that $y = (0, 0)$, $y_1 = (r, \theta)$, $y_2 = (r, \pi - \theta)$, where θ is the angle between τ and $y_1 - y$. Next, we sample y_1 according to some angular density p_θ symmetrical with respect to $n(y)$. Hence, y_2 is distributed with the same angular density. The distribution density of points y_i, $i = 1, 2$ in the measure σ is equal to

$$p_i(y, y_i) = p_\theta \cdot \frac{\cos \varphi_{y_i y}}{|y - y_i|}.$$

Therefore, we can use the following random estimator for $\partial W / \partial n(y)$:

$$\xi[W_n](y) = \frac{\partial \mu / \partial \tau(y_1)}{p_1(y, y_1)} + \frac{\partial \mu / \partial \tau(y_2)}{p_2(y, y_2)}. \tag{3.78}$$

From here, we see that the tangential derivative $\partial \mu / \partial \tau$ must be known on the boundary. According to the double randomization principle, a Monte Carlo estimator can be used instead of the exact value of this function.

If the Dirichlet BVP is considered, we have an integral equation and, hence, a random estimator for μ only. Nevertheless, correlated WB-processes starting from the points y_1, y_2 can be used here to obtain some random estimators for the tangential derivatives at these points.

4 Walk-on-boundary algorithms for the heat equation

4.1 Heat potentials and Volterra boundary integral equations

Let G be a bounded domain in $\mathbb{R}^m$, $m \geq 3$, and let the boundary of G, ∂G, be a piecewise Lyapunov-continuous surface. We denote by $Z_0(x,t)$ the fundamental solution to the heat equation:

$$Z_0(x,t) = \Theta(t)(4\pi t)^{-\frac{m}{2}} \exp\left\{-\frac{|x|^2}{4t}\right\}$$
$$= \begin{cases} (4\pi t)^{-m/2} \exp\left\{-\frac{|x|^2}{4t}\right\}, & t \geq 0,\ |x| \neq 0, \\ 0, & t < 0, \end{cases} \tag{4.1}$$

where $\Theta(t)$ is the Heaviside step function:

$$\Theta(t) = \begin{cases} 0, & t < 0 \\ 1, & t \geq 0 \end{cases}.$$

In analogy with the stationary case (Chapter 3), we introduce the surface potentials: the single-layer heat potential,

$$V(x,t) = \int_0^t d\tau \int_\Gamma 2Z_0(x-y, t-\tau)\mu(y,\tau)d\sigma(y), \tag{4.2}$$

and the double-layer heat potential,

$$W(x,t) = \int_0^t d\tau \int_\Gamma 2\frac{\partial Z_0}{\partial n_y}(x-y,t-\tau)\mu(y,\tau)d\sigma(y)$$
$$= \int_0^t d\tau \int_\Gamma \frac{|x-y|}{t-\tau} \cos\varphi_{yx} Z_0(x-y,t-\tau)\mu(y,\tau)d\sigma(y). \tag{4.3}$$

In what follows, we use the notations:

$$\begin{aligned} D_{m+1}^T &= \mathbb{R}^m \times (0,T), & Q_T &= G \times (0,T), \\ Q_T' &= G_1 \times (0,T), & S_T &= \Gamma \times [0,T], \\ G_1 &= \mathbb{R}^m \setminus \overline{G}, & \overline{G} &= G \cup \Gamma. \end{aligned} \tag{4.4}$$

If Γ is the Lyapunov surface, and $\mu \in C(S_T)$, then $V \in C(D^T_{m+1})$, and V satisfies the heat equation

$$\frac{\partial V}{\partial t} = \Delta V \tag{4.5}$$

in Q_T and Q'_T; $V(x,0) = 0$ and there exist the limiting values of normal derivatives

$$\left(\frac{\partial V}{\partial n_x}\right)_+ \quad \text{(outside, on } \Gamma\text{)},$$
$$\left(\frac{\partial V}{\partial n_x}\right)_- \quad \text{(inside, on } \Gamma\text{)}.$$

These derivatives satisfy the following equality:

$$\begin{aligned}\left(\frac{\partial V}{\partial n_x}\right)_\pm &= \pm\mu(x,t) + \frac{\partial V}{\partial n_x} \\ &= \pm\mu(x,t) + \int_0^t d\tau \int_\Gamma 2\frac{\partial Z_0}{\partial n_x}(x-y,t-\tau)\mu(y,\tau)d\sigma(y).\end{aligned} \tag{4.6}$$

Here, $\frac{\partial V}{\partial n_x}$ is the direct value of the normal derivative.

Analogously, if $\mu \in C(S_T)$, then $W \in C(\overline{Q}_T) \cup C(\overline{Q}'_T)$, and

$$\begin{aligned}W_\pm(x,t) &= \mp\mu(x,t) + W(x,t) \\ &= \mp\mu(x,t) + \int_0^t d\tau \int_\Gamma 2\frac{\partial Z_0}{\partial n_y}(x-y,t-\tau)\mu(y,\tau)d\sigma(y).\end{aligned} \tag{4.7}$$

Using (4.6), (4.7) we can reformulate BVPs for the heat equation in the form of integral equations.

Consider the interior Dirichlet problem in the domain G (or the exterior problem in the domain G_1) for the heat equation:

$$\begin{aligned}\frac{\partial u}{\partial t} &= \Delta u(x,t), \ x \in G(G_1), t > 0, \\ u(x,0) &= 0, \ x \in G(G_1), \\ u(y,t) &= g(y,t), \ y \in \Gamma, t > 0.\end{aligned} \tag{4.8}$$

We seek for the solution to (4.8) in the form of the double-layer potential W. From (4.7) we get the boundary integral equation for the density μ:

$$\mu(y,\tau) = \lambda \int_0^\tau d\tau' \int_\Gamma \frac{|y-y'|}{\tau-\tau'} \cos\varphi_{y'y} Z_0(y-y',\tau-\tau')\mu(y',\tau')d\sigma(y') + f(y,\tau), \tag{4.9}$$

where $\lambda = -1$, $f(y,\tau) = g(y,\tau)$ for the interior problem and $\lambda = 1$, $f(y,\tau) = -g(y,\tau)$ for the exterior problem.

Analogously, for the solution of the interior (exterior) Neumann problem

$$\begin{aligned} \frac{\partial u}{\partial t} &= \Delta u(x,t), \ x \in G(G_1), \ t > 0, \\ u(x,0) &= 0, \ x \in G(G_1), \\ \frac{\partial u}{\partial n_y} &= g(y,t), \ y \in \Gamma, \ t > 0, \end{aligned} \tag{4.10}$$

we use the representation in the form of the single-layer potential V.

Then (4.6) gives

$$\mu(y,\tau) = \lambda \int_0^\tau d\tau' \int_\Gamma \frac{|y-y'|}{\tau-\tau'} \cos\varphi_{yy'} Z_0(y-y',\tau-\tau')\mu(y',\tau')d\sigma(y') + f(y,\tau), \tag{4.11}$$

where $\lambda = 1, f(y,\tau) = -g(y,\tau)$ for the interior problem and $\lambda = -1, f(y,\tau) = g(y,\tau)$ for the exterior problem.

Finally, let us consider the third BVP:

$$\begin{aligned} \frac{\partial u}{\partial t} &= \Delta u(x,t), \ x \in G(G_1), \ t > 0, \\ u(x,0) &= 0, \ x \in G(G_1), \\ \frac{\partial u}{\partial n_y} - H(y,t)u(y,t) &= g(y,t), \ y \in \Gamma, \ t > 0. \end{aligned} \tag{4.12}$$

If we seek for the solution to (4.12) in the form of the single-layer potential, V, then $\mu(y,\tau)$ solves the equation

$$\begin{aligned} \mu(y,\tau) = \lambda \int_0^\tau d\tau' \int_\Gamma &\left[\frac{|y-y'|}{\tau-\tau'} \cos\varphi_{yy'} - H(y,\tau)\right] \\ &\times Z_0(y-y',\tau-\tau')\mu(y',\tau')d\sigma(y') + f(y,\tau), \end{aligned} \tag{4.13}$$

where $\lambda = 1, f = -g$ for the interior problem and $\lambda = -1, f = g$ for the exterior problem.

Note that integral equations (4.9), (4.11), (4.13) are of the Volterra' type, and their kernels include weak singularities (weak polar kernels). This implies that they have solutions for arbitrary λ, and functions μ are as smooth as the functions $g(y,\tau)$ [60]. Thus, for $g \in C(S_T)$, densities μ belong to $C(S_T)$; potentials V and W solve the heat equation and satisfy the corresponding initial and boundary conditions.

4.2 Nonstationary walk-on-boundary process

Let $X = \Gamma \times (-\infty, \infty)$. First, we assume that G is convex, and let $p_0(x,\tau)$ be a probability density in X: $p_0(y,\tau) \geq 0$, and $\int_X p_0(y,\tau)d\sigma(y)d\tau = 1$.

Definition 4.1. We call a homogeneous Markov chain $\{(y_0,\tau_0),(y_1,\tau_1),\dots\}$ defined on X the random walk-on-boundary B-process if its states are defined as follows: the point (y_0,τ_0) is distributed in accordance with a density $p_0(y,\tau)$, and the transitional density from (y,τ) to (y',τ') is

$$p(y,\tau\to y',\tau')=\frac{|y-y'|}{\tau-\tau'}\cos\varphi_{y'y}Z_0(y-y',\tau-\tau'). \tag{4.14}$$

Note that $p(y,\tau\to y',\tau')=0$ for $\tau'\ge\tau$, so in the trajectories of B-process, time is monotonically decreasing. We will use the factorized representation:

$$p(y,\tau\to y',\tau')=p_y(y\to y')p_\tau(\tau\to\tau'|y,y'), \tag{4.15}$$

where

$$p_y(y\to y')=\frac{2\cos\varphi_{y'y}}{\sigma_m|y-y'|^{m-1}} \tag{4.16}$$

is the spatial transitional density of the stationary walk-on-boundary process introduced in Chapter 3, and

$$p_\tau(\tau\to\tau'|y,y')=\frac{\pi^{m/2}|y-y'|^m}{\Gamma(\frac{m}{2})(\tau-\tau')}Z_0(y-y',\tau-\tau') \tag{4.17}$$

is the transitional density in time; $\Gamma(\cdot)$ is the Euler Γ-function.

Thus, (4.15) gives the following algorithm of simulation of the transition $(y,\tau)\to(y',\tau')$:

(a) simulate a random point $y'\in\Gamma$ using $p_y(y\to y')$, i.e., in accordance with the isotropic distribution in the solid angle of view from the point y, that is, in the same way as in the stationary walk-on-boundary process, described in Chapter 3;
(b) for y,τ,y' fixed, the parameter τ' ('time') is sampled in $(-\infty,\tau)$ using the density $p_\tau(\tau\to\tau'|y,y')$, i.e.,

$$\tau'=\tau-\frac{|y-y'|^2}{4\gamma_{m/2}}, \tag{4.18}$$

where $\gamma_{m/2}$ is γ-distributed random variable with parameter $m/2$. Its density is

$$q_{m/2}(x)=x^{m/2-1}\exp(-x)/\Gamma(m/2),\quad x>0.$$

Remark 4.2. If m is even, then the simulation formula is

$$\gamma_{m/2}=-\ln(\alpha_1\dots\alpha_{m/2}),$$

as while for odd m,

$$\gamma_{m/2}=\xi+\zeta,$$

where $\xi=\gamma_{[\frac{m}{2}]}$ and $\zeta=\eta^2/2$; η is a standard Gaussian random variable.

Definition 4.3. We define F-process of random walk on boundary as a homogeneous Markov chain on X with the initial density p_0 and the transitional density

$$p(y,\tau \to y',\tau') = \frac{|y-y'|}{\tau'-\tau} \cos\varphi_{y'y} Z_0(y-y',\tau'-\tau). \tag{4.19}$$

It is clear that $\tau' > \tau$ with probability 1. This means that time is monotonically increasing.

Simulation of the F-process:

(a) first, a point y' is sampled on Γ exactly as in the case of B-process;

(b) for fixed y, τ, y', time τ' is simulated by

$$\tau' = \tau + |y-y'|/4\gamma_{m/2},$$

which differs from (4.18) only by the sign.

The following simple statement shows the relation between B- and F-processes.

Proposition 4.4. *Let $\{(y_n,\tau_n)\}_{n=0}^{\infty}$ be B-process on Γ with the initial density $p_0(y,\tau)$, and let t be a fixed real number. Then $\{(y_n,t-\tau_n)\}_{n=0}^{\infty}$ is F-process on Γ with the initial density $p_1(y,\tau) = p_0(y,t-\tau)$.*

And vice versa, if $\{(y_n,\tau_n)\}_{n=0}^{\infty}$ is F-process on Γ with the initial density $p_0(y,\tau)$, then $\{(y_n,t-\tau_n)\}_{n=0}^{\infty}$ is B-process on Γ with the initial density $p_1(y,\tau) = p_0(y,t-\tau)$.

Proof. The proof follows from the definition of B- and F-processes and the well-known rules of transformation of distribution densities when transforming the random variables. □

Consider a non-convex domain now. In this case, the straight line

$$A_y = \{y' : y' = y + t\omega; t \in R\}, \quad \omega - \text{isotropic}$$

intersects the boundary Γ, say, in $q(y,y')$ points (not counting the point y). Then, we choose one of them at random with probability $1/q$. In this case, the transitional densities of B- and F-processes are

$$\begin{aligned} p(y,\tau \to y',\tau') &= \frac{2|\cos_{y'y}|}{\sigma_m q(y,y')|y-y'|^{m-1}} \cdot \frac{\pi^{m/2}|y-y'|^m}{\Gamma(\frac{m}{2})(\tau-\tau')} Z_0(y-y',\tau-\tau') \\ &= \frac{|\cos\varphi_{y'y}||y-y'|}{q(y,y')(\tau-\tau')} Z_0(y-y',\tau-\tau') \end{aligned} \tag{4.20}$$

and

$$p(y,\tau \to y',\tau') = \frac{|\cos\varphi_{y'y}||y-y'|}{q(y,y')(\tau'-\tau)} Z_0(y-y',\tau'-\tau), \tag{4.21}$$

respectively.

Remark 4.5. Sometimes, it is necessary to construct some modifications of B- and F-processes. In particular, we can introduce a random termination of the chain or change the transition law in space and time. For example, for the third BVP, we will define a modification of F-process introducing a non-isotropic transition.

4.3 The Dirichlet problem

As follows from the results of Section 4.2, to find the solution to interior and exterior Dirichlet problem, it is necessary to evaluate linear functional (4.3) of solutions to Volterra' integral equation (4.9). Since the Neumann series for this equation converge (the relevant spectral radius is equal to zero), we can use the standard Monte Carlo estimator described in Chapter 2. We will see that the kernel of integral equation (4.9) generates B-process.

It is convenient to extend the domain of definition of the functions μ and f in (4.9) from $\Gamma \times (0, \infty)$ to $X = \Gamma \times (-\infty, \infty)$ by putting them zero on $\Gamma \times (-\infty, 0)$. For these functions on X, we will still use the same notations, μ and f. Then, we can rewrite (4.9) in the form

$$\mu(y,\tau) = \int_X k(y,\tau;y',\tau')\mu(y',\tau')d\sigma(y')d\tau' + f(y,\tau), \tag{4.22}$$

where $k(y,\tau;y',\tau')$ is defined by

$$k(y,\tau;y',\tau') = \lambda \frac{|y-y'|}{\tau-\tau'} \cos\varphi_{y'y} Z_0(y-y',\tau-\tau'). \tag{4.23}$$

The solution to (4.8) is the linear integral functional of μ, the solution of (4.22):

$$u(x,t) = \int_X \mu(y,\tau) h_{xt}^{(1)}(y,\tau)d\sigma(y)d\tau, \tag{4.24}$$

where

$$h_{xt}^{(1)}(y,\tau) = \Theta(\tau)Z_0(x-y,t-\tau)\frac{|x-y|}{t-\tau}\cos\varphi_{yx}, \tag{4.25}$$

$$f(y,\tau) = -\lambda g(y,\tau)\Theta(\tau), \tag{4.26}$$

$\lambda = -1$ for the interior problem and $\lambda = 1$ for the exterior problem.

First, let us suppose that G is convex. Then, the kernel k in (4.23) coincides (up to a sign) with the transitional density of B-process (see formula (4.14)). Therefore, for calculating functional (4.24), it is natural to use here B-process and the adjoint estimator (Section 2.1). In this case, it is necessary to choose the initial density of

B-process so that

$$p_0(y,\tau) \neq 0 \quad \text{for} \quad 0 < \tau < t, \tag{4.27}$$

i.e., where $h^{(1)}_{xt}(y,\tau) \neq 0$.

Proposition 4.6. *The estimator*

$$\xi^{(1)}_{xt} = -Q_0 \sum_{n=0}^{\infty} \lambda^{n+1} g(y_n, \tau_n)\Theta(\tau_n) \tag{4.28}$$

is unbiased, i.e.,

$$u(x) = \mathbb{E}\xi^{(1)}_{xt}, \tag{4.29}$$

where $\{(y_n, \tau_n)\}_{n=0}^{\infty}$ is B-process with an initial density p_0 satisfying condition (4.27) and

$$Q_0 = \frac{h^{(1)}_{xt}(y_0,\tau_0)}{p_0(y_0,\tau_0)} = \frac{\Theta(\tau_0) Z_0(x-y_0, t-\tau_0)|x-y_0| \cos\varphi_{y_0x}}{p_0(y_0,\tau_0)(t-\tau_0)}. \tag{4.30}$$

Proof. The proof follows from the fact that the Neumann series for (4.22) converges and standard arguments given in Section 2.1 □

Note that for the interior problem, it is possible to simplify estimator (4.28) by choosing the initial density as

$$\begin{aligned} p_0(y,\tau) = p_0(y,\tau|x,t) &= \frac{|x-y|\cos\varphi_{yx}}{2(t-\tau)} Z_0(x-y, t-\tau) \\ &= \left[\frac{\cos\varphi_{yx}}{\sigma_m |y-x|^{m-1}}\right] \left[\frac{\pi^{m/2}|y-x|^m}{\Gamma(m/2)(t-\tau)} Z_0(x-y,t-\tau)\right]. \end{aligned} \tag{4.31}$$

To sample the first point according to this density, first simulate a point $y_0 \in \Gamma$ in the same way as in the stationary process (Chapter 3), i.e., according to the isotropic distribution in the angle of view from the point $x \in G$; then, τ_0 is found from

$$\tau_0 = t - \frac{|x-y_0|^2}{4\gamma_{m/2}}.$$

Let $N_t = \max\{n : \tau_n > 0\}$. Since $g(y_n, \tau_n) = 0$ if $\tau_n < 0$, $\lambda = -1$ and as a consequence of (4.30), (4.31), we thus get for the interior problem:

$$\xi^{(1)}_{xt} = 2\Theta(\tau_0) \sum_{n=0}^{N_t} (-1)^n g(y_n, t_n). \tag{4.32}$$

Remark 4.7. Note that in the case of the exterior problem, the initial density $p_0(y,\tau|x_0,t)$ can also be used, where $x_0 \in G$ is some auxiliary point. Note also that if it is necessary to find $u(x,t)$ for a small value of t, then it may happen that the use of (4.31) generates the initial time instances, most of which are negative. Therefore, it is better to take

$$p_0(y,\tau) = \frac{\cos\varphi_{yx}}{\sigma_m|y-x|^{m-1}} p_t(\tau), \tag{4.33}$$

where $p_t(\tau)$ is a suitable density on $(0,t)$.

Consider now non-convex G. In this case, we use B-process $\{(y_n,\tau_n)\}_{n=0}^{\infty}$ with an initial density, which is consistent with the function $h_{xt}^{(1)}(y,\tau)$, while the transitional density is taken from (4.20). The unbiased estimator for $u(x,t)$ is

$$\xi_{xt}^{(1)} = -\sum_{n=0}^{\infty} Q_n \lambda^{n+1} g(y_n,\tau_n)\Theta(\tau_n), \tag{4.34}$$

where $\lambda = -1$ for the interior problem and $\lambda = 1$ for the exterior problem, and

$$Q_0 = \frac{h_{xt}^{(1)}(y_0,\tau_0)}{p_0(y_0,\tau_0)} = \frac{\Theta(\tau_0)Z_0(x-y,t-\tau_0)|x-y_0|\cos\varphi_{y_0x}}{(t-\tau_0)p_0(y_0,\tau_0)}, \tag{4.35}$$

$$Q_n = Q_{n-1}\cdot \operatorname{sign}\{\cos\varphi_{y_ny_{n-1}}\} q(y_{n-1},y_n), \quad n = 1,2,\ldots. \tag{4.36}$$

Again, since $g(y_n,\tau_n) = 0$ for $\tau_n < 0$, we get

$$\xi_{xt}^{(1)} = -\sum_{n=0}^{N_t} Q_n \lambda^{n+1} g(y_n,\tau_n), \tag{4.37}$$

where $N_t = \max\{n : \tau_n > 0\}$.

Remark 4.8. To find the solution at several points (x_i,t_i), $i = 1,2,\ldots,k$, we do not need to construct k Markov chains. In fact, it is sufficient to use only one B- (or F-) process. We choose the initial density $p_0(y,\tau)$ consistent with every function $h_{x_i,t_i}(y,\tau)$, $i = 1,\ldots,k$, i.e., $p_0(y,\tau) \neq 0$ for $0 < \tau < \max_i t_i$, and $\cos\varphi_{yx_i} \neq 0$. Then,

$$\xi_{x_it_i}^{(1)} = Q_0^{(i)} \sum_{n=0}^{N_t} Q_n' \lambda^{n+1} g(y_n,\tau_n), \quad i = 1,2,\ldots,k \tag{4.38}$$

are unbiased estimators for $u(x_i,t_i)$, $i = 1,2,\ldots,k$. Here,

$$Q_0^{(i)} = \frac{h_{x_it_i}^{(1)}(y_0,\tau_0)}{p_0(y_0,\tau_0)} = \frac{\Theta(\tau_0)Z_0(x_i-y_0,t_i-\tau_0)|x_i-y_0|\cos\varphi_{y_0x_i}}{(t_i-\tau_0)p_0(y_0,\tau_0)}, \tag{4.39}$$

$$Q_0' = 1,\ Q_n' = Q_{n-1}' \operatorname{sign}\{\cos\varphi_{y_ny_{n-1}}\} q(y_{n-1},y_n), \quad n = 1,2,\ldots.$$

Note also that it is easy to write down estimators for the derivatives $D_x^\alpha D_t^\beta u(x,t)$. These estimators have the same form as (4.37). The difference is in the initial weight:

$$Q_0 = \frac{D_x^\alpha D_t^\beta h_{xt}^{(1)}(y_0,\tau_0)}{p_0(y_0,\tau_0)}.$$

4.4 The Neumann problem

In this section, for functions of (y,τ) defined on $\Gamma \times [0,T]$, we use the same extensions as in Section 4.3. Such approach makes possible rewriting (4.11) in the following form:

$$\mu(y,\tau) = \int_X k(y,\tau;y',\tau')\mu(y',\tau')d\sigma(y')d\tau' + f(y,\tau). \tag{4.40}$$

Then the solution to (4.10) is written as the linear functional of μ

$$u(x,t) = \int_X \mu(y,\tau)h_{xt}^{(2)}(y,\tau)d\sigma(y)d\tau, \tag{4.41}$$

where

$$h_{xt}^{(2)} = 2\Theta(\tau)Z_0(x-y,t-\tau), \tag{4.42}$$

$$k(y,\tau;y',\tau') = \lambda\frac{|y-y'|}{\tau-\tau'}\cos\varphi_{yy'}Z_0(y-y',\tau-\tau'), \tag{4.43}$$

$$f(y,\tau) = -\lambda\Theta(\tau)g(y,\tau), \tag{4.44}$$

where $\lambda = 1$ for the interior problem and $\lambda = -1$ for the exterior Neumann problem. Note that, compared to (4.23), there is an inversion $y \leftrightarrow y'$ in the argument of the cos-function.

Therefore, it is convenient to choose the direct Monte Carlo estimator and use F-process. Consequently, the initial density $p_0(y,\tau)$ has to be chosen consistent with $f(y,\tau)$, i.e., $p_0(y,\tau) \neq 0$ when $\tau > 0$ and $g(y,\tau) \neq 0$.

Proposition 4.9. *The random variable*

$$\begin{aligned}\xi_{xt}^{(2)} &= -\sum_{n=0}^{\infty}\lambda^{n+1}Q_n h_{xt}^{(2)}(y_n,\tau_n)\\ &= -2\sum_{n=0}^{\infty}\lambda^{n+1}Q_n\Theta(\tau_n)Z_0(x-y_n,t-\tau_n),\end{aligned} \tag{4.45}$$

where

$$Q_0 = \Theta(\tau_0)g(y_0,\tau_0)/p_0(y_0,\tau_0),$$
$$Q_n = Q_{n-1}\mathrm{sign}\{\cos\varphi_{y_n y_{n-1}}\}q(y_n,y_{n-1}), \quad n=1,2,\ldots, \tag{4.46}$$

is an unbiased estimator for the solution to the Neumann problem. □

Proof. The proof follows by standard arguments, if we take into account that the transitional density of F-process $\{(y_n,\tau_n)\}_{n=0}^{\infty}$ has form (4.21). □

Note that since $Z_0(x-y_n,t-\tau_n)=0$ for $\tau_n>\tau$, we can rewrite (4.45) in the following form:

$$\xi_{xt}^{(2)} = -2\sum_{n=0}^{N_t}\lambda^{n+1}Q_n\Theta(\tau_n)Z_0(x-y_n,t-\tau_n), \tag{4.47}$$

where

$$N_t = \max\{n:\tau_n<t\}.$$

Remark 4.10. To find the solution at several points (x_i,t_i), $i=1,2,\ldots,k$, we can use only one F-process, as described in Remark 4.8. The correlated estimators for $u(x_i,t_i)$ have the form

$$\xi_{x_it_i}^{(2)} = -2\sum_{n=0}^{\infty}\lambda^{n+1}Q_n\Theta(\tau_n)Z_0(x_i-y_n,t_i-\tau_n), \tag{4.48}$$

where the random weights Q_n are defined by (4.46), and the summation in (4.48) is taken over $n:\tau_n<t_i$.

For $x\in G$, the unbiased estimator for the derivative $D_x^{\alpha}D_t^{\beta}u(x,t)$ is

$$\bar{\xi}_{xt}^{(2)} = -2\sum_{n=0}^{N_t}\lambda^{n+1}Q_n\Theta(\tau_n)D_x^{\alpha}D_t^{\beta}Z_0(x-y_n,t-\tau_n),$$

where $\lambda=1$ for the interior problem and $\lambda=-1$ for the exterior Neumann problem.

4.5 Third BVP

We proceed in the same way as in Sections 4.3 and 4.4 and rewrite boundary integral equation (4.13) in the form:

$$\mu(y,\tau) = \int_X k(y,\tau;y',\tau')\mu(y',\tau')d\sigma(y')d\tau' + f(y,\tau). \tag{4.49}$$

Then the solution to the third BVP is given by

$$u(x,t) = \int_X \mu(y,\tau) h_{xt}^{(3)}(y,\tau) d\sigma(y) d\tau. \tag{4.50}$$

Here,

$$h_{xt}^{(3)}(y,\tau) = h_{xt}^{(2)}(y,\tau) = 2\Theta(\tau) Z_0(x-y, t-\tau),$$

$$k(y,\tau;y',\tau') = \lambda\left\{\frac{|y-y'|}{\tau-\tau'}\cos\varphi_{yy'} - H(y,\tau)\right\} Z_0(y-y', \tau-\tau'), \tag{4.51}$$

$$f(y,\tau) = -\lambda\Theta(\tau) g(y,\tau), \tag{4.52}$$

with $\lambda = 1$ for the interior problem and $\lambda = -1$ for the exterior problem.

We use F-process $\{(y_n, \tau_n)\}_{n=0}^{\infty}$ with an initial density $p_0(y,\tau)$ consistent with $f(y,\tau)$.

Let

$$\xi_{xt}^{(3)} = -\sum_{n=0}^{N_t} 2\lambda^{n+1} Q_n \Theta(\tau_n) Z_0(x-y_n, t-\tau_n), \tag{4.53}$$

where

$$N_t = \max\{n : \tau_n < t\},$$

$$Q_0 = \Theta(\tau_0) g(y_0, \tau_0)/p_0(y_0, \tau_0),$$

$$Q_n = Q_{n-1}\left\{1 - \frac{H(y_n, \tau_n)(\tau_n - \tau_{n-1})}{|y_n - y_{n-1}|\cos\varphi_{y_n y_{n-1}}}\right\}.$$

In the next section, we will prove that the Neumann series for the integral equation with the kernel $|k|$ converges. Here, k is given by (4.51). Consequently, $\xi_{xt}^{(3)}$ is an unbiased estimator for functional (4.50). The variance of this estimator can be finite or infinite. For example, if the curvature of the boundary is arbitrary small at a point, then the variance is infinite. In this case, it is possible to construct a modified walk-on-boundary process. Indeed, let us describe such a modification. First, consider the 3D case. Let x_0 be a point in G such that

$$\inf_{y\in\Gamma} \cos\varphi_{yx_0} > 0.$$

Let $\omega_0 = (y' - x_0)/|y' - x_0|$, and let $p(\omega|\omega_0)$ be a density (on a unit sphere) with a singularity of the type $|\omega - \omega_0|^{-\beta} (0 < \beta < 1)$ at the direction ω_0. We can construct the random distribution as follows. Choose the coordinate system in such a way that ω_0 coincides with the x_3 axis. Then, if ω is isotropic, the latitude angle φ is uniformly distributed on $[0, 2\pi]$, and the altitude angle ψ is distributed with the density $\sin\psi$ on $[0, \pi]$. This density behaves like $|\omega - \omega_0|$ at the direction ω_0.

If we choose ψ with the distribution density $c_\beta \psi^{-\beta}$ on $[0, \pi]$, where

$$c_\beta = \left[\int_0^\pi t^{-\beta} dt \right]^{-1},$$

then the density $p(\omega|\omega_0)$ has a singularity $|\omega - \omega_0|^{-\beta}$. Thus, the density

$$\frac{\cos \varphi_{yx_0}}{|y - x_0|^2} p(\omega|\omega_0)$$

has the singularity $|y - y'|^{-\beta}$, i.e.,

$$\frac{C_1}{|y - y'|^\beta} \le \frac{\cos \varphi_{yx_0}}{|y - x_0|^2} p(\omega|\omega_0) \le \frac{C_2}{|y - y'|^\beta}, \quad C_1, C_2 > 0.$$

Consider the distribution density of the transition $(y', \tau') \to (y, \tau)$:

$$p_3(y', \tau' \to y, \tau) = \frac{\cos \varphi_{yx_0} \cdot p(\omega|\omega_0)}{|y - x_0|^2} \times \left[\frac{|y - y'|}{\sqrt{\pi}(4(\tau - \tau'))^{1/2}(\tau - \tau')} \exp\left\{ -\frac{|y - y'|^2}{4(\tau - \tau')} \right\} \right]. \tag{4.54}$$

Thus, to simulate $(y', \tau') \to (y, \tau)$, we first sample the direction ω in accordance with $p(\omega|\omega_0)$ and take the next point in space, y, as the point of intersection of a ray, starting from x_0 and having the direction ω, with the boundary Γ. In this case, y is distributed with the density

$$\frac{\cos \varphi_{yx_0} \cdot p(\omega|\omega_0)}{|y - x_0|^2}.$$

The next instant of time is simulated using the formula

$$\tau = \tau' + |y - y'|^2 / 4\gamma_{1/2},$$

where $\gamma_{1/2}$ is a sample value of the γ-distributed random variable with the parameter equal to $1/2$. This implies that τ is distributed on (τ', ∞) with the following density:

$$p(\tau|\tau', y, y') = \frac{|y - y'| \exp\left\{ -\frac{|y - y'|^2}{4(\tau - \tau')} \right\}}{\sqrt{\pi}(\tau - \tau')[4(\tau - \tau')]^{1/2}}.$$

In two dimensions, we can use the density

$$p_2(y', \tau' \to y, \tau) = \frac{\cos \varphi_{yx_0} \cdot p(\omega|\omega_0)}{|y - x_0|} \times \left[\frac{|y - y'|^\beta \exp\{-|y - y'|^2 / 4(\tau - \tau')\}}{\Gamma(\beta/2)(\tau - \tau')(4(\tau - \tau'))^{\beta/2}} \right], \tag{4.55}$$

where $p(\omega|\omega_0)$ is a density on the unit circle, having a singularity of the type $|\omega-\omega_0|^{-\beta}$. The spatial sampling is the same as in the 3D case, and the next instant of time is taken as

$$\tau = \tau' + |y-y'|^2/4\gamma_{\frac{\beta}{2}},$$

where $\gamma_{\beta/2}$ is a sample value of the γ-distributed random variable with the parameter equal to $\beta/2$.

Let $\{(\bar{y}_n, \bar{\tau}_n)\}_{n=0}^{\infty}$ be a homogeneous Markov chain in a phase space $X = \Gamma \times (-\infty, \infty)$ with an initial density $p_0(y, \tau)$ consistent with function (4.52) and with transitional density (4.54) (3D) or (4.55) (2D). Then, the random variable

$$\xi_{xt}^{(3)} = -\sum_{n=0}^{N_t} 2\bar{Q}_n \Theta(\bar{\tau}_n) Z_0(x-\bar{y}_n, t-\bar{\tau}_n) \tag{4.56}$$

is an unbiased estimator for $u(x,t)$, the solution to the third BVP. In (4.56),

$$\begin{aligned}
\bar{Q}_0 &= \frac{\Theta(\bar{\tau}_0) g(\bar{y}_0, \bar{\tau}_0)}{p_0(\bar{y}_0, \bar{\tau}_0)}, \\
\bar{Q}_n &= \bar{Q}_{n-1} \left\{ \frac{|\bar{y}_n - \bar{y}_{n-1}|}{\bar{\tau}_n - \bar{\tau}_{n-1}} \cos\varphi_{\bar{y}_n\bar{y}_{n-1}} - H(\bar{y}_n, \bar{\tau}_n) \right\} \\
&\quad \times \frac{|\bar{y}_n - x_0|^2}{4\pi |\bar{y}_n - \bar{y}_{n-1}| \cos\varphi_{\bar{y}_n x_0} p(\omega_n|\omega_{n-1}^*)}
\end{aligned}$$

in 3D case, and

$$\begin{aligned}
\bar{Q}_n &= \bar{Q}_{n-1} \left\{ \frac{|\bar{y}_n - \bar{y}_{n-1}|}{\bar{\tau}_n - \bar{\tau}_{n-1}} \cos\varphi_{\bar{y}_n\bar{y}_{n-1}} - H(\bar{y}_n, \bar{\tau}_n) \right\} \\
&\quad \times \frac{|\bar{y}_n - x_0| \Gamma(\beta/2)(4\pi(\bar{\tau}_n - \bar{\tau}_{n-1}))^{\beta/2}}{4\pi |\bar{y}_n - \bar{y}_{n-1}| \cos\varphi_{\bar{y}_n x_0} p(\omega_n|\omega_{n-1}^*)}
\end{aligned}$$

in two dimensions, where

$$\omega_{n-1}^* = (\bar{y}_{n-1} - x_0)/|\bar{y}_{n-1} - x_0|.$$

We analyse unbiasedness and variance of the estimator (4.56) in the next section.

Remark 4.11. It is possible to weaken the condition

$$\inf_{y\in\Gamma} \cos\varphi_{yx_0} > 0.$$

Suppose that there exist points $x_0, x_1, \dots, x_l \in G$ such that

$$\inf_{y\in\Gamma} \sum_{i=0}^{l} |\cos\varphi_{yx_i}| > 0.$$

To construct a Markov chain in this case, we can use a mixture (a linear combination) of transitional densities of type (4.54) in 3D or (4.55) in 2D.

4.6 Unbiasedness and variance of the walk-on-boundary algorithms

In this section, we give a rigorous proof that all the estimators constructed in Chapter 4 are unbiased and have a finite variance. Also we estimate the cost of the walk-on-boundary algorithms in application to solving BVPs for the heat equation.

All the Monte Carlo estimators constructed in this chapter are based on a reformulation of the original problem in the form of a Volterra–Fredholm integral equation of the second kind:

$$\mu(y,\tau) = \int_0^{\tau} d\tau' \int_{\partial G} k(y,\tau;y',\tau')\mu(y',\tau')d\sigma(y') + f(y,\tau). \tag{4.57}$$

For arbitrary $(y,\tau) \in \Gamma_t \equiv \partial G \times (0,t)$ and $(y',\tau') \in \Gamma_t$, the kernels $k(y,\tau;y',\tau')$ of these equations satisfy the conditions

$$|k(y,\tau;y',\tau')| \le c[\tau-\tau]^{-(m/2+\alpha)}|y-y'|^{\beta}\exp\left\{-\frac{\gamma|y-y'|^2}{\tau-\tau'}\right\}. \tag{4.58}$$

Generally, positive constants c and γ in (4.58) may depend on t, and α, β are some parameters.

Denote by $K_{\alpha\beta}$ the class of kernels satisfying this inequality. To investigate the variance and the cost of algorithms presented in this chapter, we first derive some properties of solutions to equations of type (4.58) whose kernels belong to $K_{\alpha\beta}$.

Lemma 4.12. *If $k \in K_{\alpha\beta}$, then $k \in K_{\alpha-\varepsilon/2,\beta-\varepsilon}$ for any $\varepsilon > 0$.*

Proof. Indeed,

$$\left[|y-y'|/(\tau-\tau')^{1/2}\right]^{\varepsilon}\exp\left\{-\frac{\delta|y-y'|^2}{\tau-\tau'}\right\} < \infty$$

since the function $x^{\varepsilon}\exp\{-\delta x\}$ is bounded on $(0,\infty)$ for arbitrary $\delta > 0$. Consequently, from $k \in K_{\alpha\beta}$, we get

$$\begin{aligned}|k(y,\tau;y',\tau')| &\le c[\tau-\tau']^{-(m/2+\alpha)}|y-y'|^{\beta}\exp\left\{-\frac{\gamma|y-y'|^2}{\tau-\tau'}\right\}\\ &\le c_1[\tau-\tau']^{-(m/2+\alpha-\varepsilon/2)}|y-y'|^{\beta-\varepsilon}\exp\left\{-\frac{\gamma|y-y'|^2}{2(\tau-\tau')}\right\}.\end{aligned}$$

□

This property implies that it is possible to simplify the two-parameter expressions depending on α and β by passing to a one-parameter formulation.

Lemma 4.13. *If $k \in K_{\alpha\beta}$ where*

$$\delta = \frac{1}{2}(\beta - 2\alpha + 1) > 0, \quad \beta + m - 1 > 0, \tag{4.59}$$

and the free term of (4.57) satisfies the condition

$$|f(y,\tau)| \le c_f \tau^a, \quad (y,\tau) \in \Gamma_t, \quad a > -1, \tag{4.60}$$

where c_f does not depend on y and τ, and ∂G is a Lyapunov surface, then the Neumann series for (4.57) converges and

$$|[K^j f](y,\tau)| \le c c_f \Gamma(a+1)\tau^a \frac{(A\tau^\delta)^j}{\Gamma(a+1+j\delta)}, \tag{4.61}$$

where A is a constant, which does not depend on α, β and ∂G.

Proof. We prove this property by using some general results presented in [89]. Denote by $\omega_y(r)$ the surface area of a part of ∂G that lies inside a ball $B(x,r)$:

$$\omega_y(r) = \int_{|y-y'|<r} d\sigma(y').$$

Then

$$\int_{\partial G} f(|y-y'|) d\sigma(y') = \int_0^\infty f(r) d\omega_y(r),$$

and

$$|[Kf](y,\tau)| \le c c_f \int_0^\tau d\tau' \int_{\partial G} |y-y'|^\beta \exp\left\{-\frac{\gamma |y-y'|^2}{\tau-\tau'}\right\} \frac{\tau'^a d\sigma(y')}{[\tau-\tau']^{m/2+\alpha}}$$

$$= c c_f \int_0^\tau d\tau' \int_{\partial G} \frac{r^\beta \tau'^a}{[\tau-\tau']^{m/2+\alpha}} \exp\left\{-\frac{\gamma r^2}{\tau-\tau'}\right\} d\omega_y(r). \tag{4.62}$$

Since ∂G is a Lyapunov surface, there exist such constants $0 < b < B$ that $br^{m-1} < \omega_y(r) < Br^{m-1}$ and therefore

$$[\omega_y(r)B^{-1}]^{\frac{1}{m-1}} \le r \le [\omega_y(r)b^{-1}]^{\frac{1}{m-1}}. \tag{4.63}$$

From this we get

$$|[Kf](y,\tau)| \le c_f c \int_0^\tau d\tau' \int_0^\infty (\tau-\tau')^{-m/2-\alpha} [\omega_y(r)/b]^{\frac{\beta}{m-1}}$$

$$\times \tau'^a \exp\left\{-\gamma \frac{(\omega_y(r)/B)^{\frac{2}{m-1}}}{\tau-\tau'}\right\} d\omega_y(r).$$

Using a new variable $\varrho = \gamma[\omega_y(r)B^{-1}]^{\frac{2}{m-1}}/(\tau - \tau')$ we transform the right-hand side to

$$c_f C(b, B, \beta, \gamma, c) \int_0^\tau d\tau'(\tau - \tau')^{\delta-1}\tau'^a \int_0^\infty \varrho^{(\beta+m-3)/2} \exp\{-\varrho\} d\varrho,$$

and hence

$$|[Kf](y, \tau)| \le c_f CB(a + 1, \delta)\tau^{a+\delta}, \tag{4.64}$$

where $B(.,.)$ is the Euler beta-function. Recurrent application of (4.64) gives

$$|[K^j f](y, \tau)| \le c_f C^j B(a + 1, \delta) \dots B(a + 1 + (j - 1)\delta, \delta)\tau^{a+j\delta}, \quad j = 2, 3, \dots .$$

Substitution of the beta-function representation in terms of Γ-function, $B(x, y) = \Gamma(x)\Gamma(y)/\Gamma(x + y)$, yields

$$\begin{aligned} |[K^j f](y, \tau)| &\le c_f C^j \frac{\Gamma(a + 1)\Gamma(\delta)}{\Gamma(a + 1 + \delta)} \frac{\Gamma(a + 1 + \delta)\Gamma(\delta)}{\Gamma(a + 1 + 2\delta)} \dots \\ &\dots \frac{\Gamma(a + 1 + (j - 1)\delta)\Gamma(\delta)}{\Gamma(a + 1 + j\delta)} \tau^{a+j\delta} \\ &= c_f \Gamma(a + 1)\tau^a \frac{(A\tau^\delta)^j}{\Gamma(a + 1 + j\delta)} \equiv r_j, \end{aligned}$$

where $A = C\Gamma(\delta)$. Thus, the Neumann series converges, since

$$\sum_{j=0}^\infty r_j = \sum_{j=0}^\infty \frac{a_0^j}{\Gamma(a + 1 + j\delta)} < \infty$$

for arbitrary positive $a_0 = A\tau^\delta$ and δ, due to

$$\frac{r_{j+1}}{r_j} = C\tau^\delta B(a + 1 + j\delta, \delta) \to 0,$$

when $j \to \infty$. □

Definition 4.14. Kernels k that belong to a class $K_{\alpha\beta}$ and satisfy the conditions (4.59) are called weakly polar.

Lemma 4.15. *Let*

$$k(y, \tau; y', \tau') = \sum_{i=1}^n k_i(y, \tau; y', \tau').$$

Suppose that $k_i \in K_{\alpha_i,\beta_i}$, $i = 1, \dots, n$, and all k_i are weakly polar kernels, i.e., condition (4.58) holds for some constants c_i, γ_i. Assume also that $|f(y, \tau)| \le c_f$, $(y, \tau) \in \Gamma_t$. Then the Neumann series for (4.57) converges.

Proof. Indeed, from (4.64), we have ($a = 0$)

$$|[Kf](y,\tau)| \le \sum_{i=1}^{n} |[K_i f](y,\tau)| \le \sum_{i=1}^{n} c_f B(1,\delta_i) C_i \tau^{\delta_i},$$

where K_i is the integral operator generated by the kernel k_i and $\delta_i = (\beta_i - 2\alpha_i + 1)/2 > 0$, $i = 1, 2, \dots, n$. From this, we get

$$\begin{aligned}|[K^j f](y,\tau)| &\le \sum_{i_1=1}^{n} \dots \sum_{i_j=1}^{n} c_f C_{i_1} \dots C_{i_j} B(1,\delta_{i_1}) B(1+\delta_{i_1},\delta_{i_2}) \\ &\quad \dots B(1+\delta_{i_1}+\dots+\delta_{i_{j-1}},\delta_{i_j}) \tau^{\delta_{i_1}+\delta_{i_2}+\dots+\delta_{i_j}} \\ &= \sum_{i_1=1}^{n} \dots \sum_{i_j=1}^{n} c_f C_{i_1} \dots C_{i_j} \frac{\Gamma(1)\Gamma(\delta_{i_1})\dots\Gamma(\delta_{i_j})}{\Gamma(1+\delta_{i_1}+\dots\delta_{i_j})} \tau^{\delta_{i_1}+\delta_{i_2}+\dots+\delta_{i_j}}.\end{aligned}$$

Hence

$$|[K^j f](y,\tau)| \le c_f (nC_{\max}\gamma)^j \frac{\tau^{j\delta^0}}{\Gamma(1+j\delta_{\min})},$$

where $\delta_{\min} = \min_i \delta_i$, $C_{\max} = \max_i C_i$, $\delta^0 = \delta_{\min}$ if $0 < \tau \le 1$ and $\delta^0 = \delta_{\max} = \max_i \delta_i$ if $\tau > 1$; $\gamma = \max_i \Gamma(\delta_i)$. This completes the proof. □

We now turn to the estimators $\eta_{xt}^{(i)}$, $i = 1, 2, 3, 4$, and prove that they are unbiased and have finite variances.

Unbiasedness of these estimators follows from the convergence of Neumann series for (4.9), (4.11), (4.13). Indeed, $|\cos\varphi| < c|y - y'|^\lambda$, where φ equals to $\varphi_{yy'}$ or to $\varphi_{yy'}$, $0 < \lambda \le 1$, since ∂G is a Lyapunov surface. Hence, the kernels of (4.9), (4.11) belong to $K_{1,1+\lambda}$ (weakly polar kernels), and from Lemma 4.2 we conclude that the Neumann series converges if the right-hand side is bounded. Next, the kernel of (4.13) is represented as a sum of two kernels: the first kernel belongs to $K_{1,1+\lambda}$ and the second one lies in $K_{0,0}$. Thus, by Lemma 4.3, the Neumann series for (4.13) converges.

To prove that the variance of $\eta_{xt}^{(1)}$ is finite, it is sufficient to show that the Neumann series for the equation

$$\varphi^*(y,\tau) = \int_0^\tau d\tau' \int_{\partial G} \frac{k^2(y,\tau;y',\tau')}{p(y,\tau \to y',\tau')} \varphi^*(y',\tau') d\sigma(y') + \frac{\left(h_{xt}^{(1)}(y,\tau)\right)^2}{p_0(y,\tau)}$$

converges. Here, $k(y,\tau;y',\tau')$ is the kernel of (4.9) and p is the transitional density of B-process. Convergence of the Neumann series follows now from the fact that $k^2/p \le$ const $|k|$, and so $k^2/p \in K_{1,1+\lambda}$. Clearly, p_0 must be consistent with $h_{xt}^{(1)}$.

For the estimator $\eta_{xt}^{(2)}$, the proof is fully analogous.

To prove that the variance of $\eta_{xt}^{(3)}$ is finite, it is sufficient to prove that the Neumann series for the equation

$$\varphi^*(y,\tau) = \int_0^\tau d\tau' \int_{\partial G} \frac{k^2(y,\tau;y',\tau')}{p_3(y',\tau' \to y,\tau)} \varphi^*(y',\tau') d\sigma(y') + \frac{f^2(y,\tau)}{p_0(y,\tau)}$$

converges. Here, $k(y,\tau;y',\tau')$ is the kernel of (4.13). We have

$$k = k_1 + k_2 + k_3,$$

where

$$\begin{aligned} k_1 &= \frac{|y-y'|}{\tau-\tau'} \cos\varphi_{yy'} Z_0(y-y',\tau-\tau'), \\ k_2 &= -4H(y,\tau) Z_0(y-y',\tau-\tau'), \\ k_3 &= \frac{4H^2(y,\tau)(\tau-\tau')}{|y-y'|\cos\varphi_{yy'}} Z_0(y-y',\tau-\tau'). \end{aligned}$$

Note that k_1 and k_2 are weakly polar kernels, while k_3 becomes weakly polar if additional assumptions are made. Indeed, if $\cos\varphi_{yy'} \sim |y-y'|$, then $k_3 \in K_{-1,-2}$, and hence the condition (4.59) is not satisfied. So, we assume that the boundary ∂G satisfies the following condition:

$$\cos\varphi_{yy'} < c|y-y'|^{\lambda_1}, \quad y,y' \in \partial G, \quad 1 > \lambda_1 > 0.$$

Then, $k_3 \in K_{-1,-1-\lambda_1}$, and therefore the kernel k_3 is weakly polar. Thus, for such boundaries, the variance of $\eta_{xt}^{(3)}$ is finite.

Now we show that the variance of $\eta_{xt}^{(4)}$ is finite if the Lyapunov parameter λ is larger than 1/2. It suffices to prove that k^2/p_4 is a weakly polar kernel, where k is the kernel of (4.13). Here, we have

$$\begin{aligned} c_1[\tau-\tau']^{-3/2}|y-y'|^{1-\beta} \exp\left\{-\frac{|y-y'|^2}{4(\tau-\tau')}\right\} &\le p_4(y',\tau' \to y,\tau) \\ \le c_2[\tau-\tau']^{-3/2}|y-y'|^{1-\beta} \exp\left\{-\frac{|y-y'|^2}{4(\tau-\tau')}\right\}&, \end{aligned}$$

and therefore

$$k^2/p_4 = k_1(y',\tau';y,\tau) + k_2(y',\tau';y,\tau) + k_3(y',\tau';y,\tau),$$

where $k_1 \in K_{2,1+2\lambda+\beta}$, $k_2 \in K_{1,1+\beta}$, $k_3 \in K_{0,\beta-1}$. Consequently, the kernels k_1, k_2 and k_3 are all weakly polar if

$$2\lambda + \beta > 2, \quad \lambda + \beta > 1. \tag{4.65}$$

The second inequality in (4.65) can be satisfied for arbitrary $\lambda > 0$ by an appropriate choice of $\beta \in (0,1)$, while the first inequality is satisfied for $\lambda > 1/2$.

4.7 The cost of the walk-on-boundary algorithms

First, we consider the walk-on-boundary algorithms for solving the Neumann problem. For this problem, the solution is based on simulation of F-process, $\{(y_n, \tau_n)\}_{n=0}^{\infty}$. The estimator cost is proportional to the average number of transitions in the Markov chain. Thus, we need to estimate $\mathbb{E}N_t$.

Proposition 4.16. *Suppose that the boundary Γ satisfies the following condition (interior cone condition): there exist such two numbers $\alpha_0 \geq 0$ and $h_0 > 0$ that for an arbitrary $y \in \Gamma$, the cone $K_y(\alpha_0, h_0)$ lies inside G.*
Here, the cone is defined by its vertex, y, angle, α_0, and height, h_0, and both angle and height do not depend on y.

Suppose that the initial density of F-process is taken so that $\tau_0 > 0$ with probability 1. Then, the following asymptotic inequality is true:

$$\mathbb{E}N_t \leq ct$$

for large t. Constant value, c, depends on α_0 and h_0.

Proof. Let

$$\Delta\tau_i = |y_i - y_{i-1}|^2/4\gamma_{m/2}^{(i)}.$$

Then,

$$N_t = \max\{n : \tau_0 + \Delta\tau_1 + \cdots + \Delta\tau_n < t\}.$$

Denote $\rho_i = |y_i - y_{i-1}|$, and let

$$\rho_i' = \begin{cases} h_0, & \text{if the vector } y_i - y_{i-1} \text{ belongs to } K_{y_{i-1}}(\alpha_0, h_0) \\ 0, & \text{otherwise} \end{cases}.$$

Then, $\rho_i \geq \rho_i'$; ρ_i' are independent, and $\rho_i' = h_0$ with probability $2\alpha_0/\sigma_m$, while $\rho_i' = 0$ with probability $1 - 2\alpha_0/\sigma_m$.

Let δ be a fixed positive number. We introduce a δ-cut for the sequence $\gamma_{m/2}^{(i)}$ by the following definition:

$$\gamma_i' = \begin{cases} \gamma_{m/2}^{(i)}, & \gamma_{m/2}^{(i)} \geq \delta \\ \delta, & \gamma_{m/2}^{(i)} < \delta \end{cases}.$$

Let

$$\Delta\tau_i' = (\rho_i')^2/4\gamma_i', \quad i = 1, 2, \ldots.$$

Since $\Delta\tau_i \geq \Delta\tau_i'$, and τ_0 is positive with probability 1, we conclude that the inequality

$$N_t \leq N_t' = \max\{n : \Delta\tau_1' + \Delta\tau_2' + \cdots + \Delta\tau_n' < t\}$$

holds with probability 1. The random variables $\Delta\tau_1', \Delta\tau_2', \ldots$ are independent, equally distributed with a finite expectation μ and variance σ^2. Consequently, by the renewal theorem [13], the following asymptotic relation is true:

$$\mathbb{E}N_t' = \frac{t}{\mu} + O(t), \; \mathbb{D}N_t' = \frac{\sigma^2 t}{\mu^3} + O(t) \tag{4.66}$$

as $t \to \infty$.

Therefore, from $N_t' \geq N_t$, it follows that

$$\mathbb{E}N_t \leq C_1 t, \quad \mathbb{D}N_t \leq C_2 t^2$$

with probability 1.

Consider now a B-process $\{(y_n, \tau_n)\}_{n=0}^{\infty}$ with the probability density of the initial point, $p_0(y, \tau)$. We suppose that $\tau_0 \leq t$ with probability 1 (this is consistent with the condition $p_0(y, \tau) \neq 0$, $0 < \tau < t$). Let $N_t = \max\{n : \tau_n > 0\}$. Therefore, if the interior cone condition is satisfied, then (4.66) is true. The proof is the same as we presented for the F-process. □

The estimations we obtained can be used for estimating the variance of $\xi_{xt}^{(i)}$. Indeed, let us consider, for example, the estimator $\xi_{xt}^{(1)}$.

Denote by $d(x)$ the distance from x to the boundary Γ:

$$d(x) = \inf\{|x - y| : y \in \Gamma\}.$$

Then, if $p_0(y, \tau) > C_3 > 0$ for all (y, τ), for which $h_{xt}^{(1)}(y, \tau) \neq 0$, we get the following estimate:

$$|Q_0| = \left| \frac{h_{xt}^{(1)}(y_0, \tau_0)}{p_0(y_0, \tau_0)} \right| \leq \frac{C_4}{(d(x))^{m+1}}. \tag{4.67}$$

If the function $g(y, \tau)$ is bounded, then using the representation for $\xi_{xt}^{(1)}$, we get from (4.66) and (4.67),

$$\mathbb{E}(\xi_{xt}^{(1)})^2 \leq \frac{C_5 t^2}{(d(x))^{2(m+1)}}.$$

The same result can be proved for the estimator $\xi_{xt}^{(2)}$ in the case of bounded $g(y,\tau)$ and $p_0 > C_6 > 0$. Thus, the cost of the walk-on-boundary algorithms behaves asymptotically like Ct^3 as $t \to \infty$.

4.8 Inhomogeneous heat equation

Consider the inhomogeneous heat equation,

$$\frac{\partial u}{\partial t} = \Delta u(x,t) + F(x,t), \; x \in G(G_1), \; 0 < t < T, \tag{4.68}$$

subject to initial condition

$$u(x,0) = \varphi(x), \quad x \in G(G_1), \tag{4.69}$$

and boundary condition

$$Bu|_\Gamma = g(y,t), \tag{4.70}$$

where B is the boundary operator of the first, second or third kind. We suppose that the function F is Hölder continuous with respect to the arguments x,t and that φ is continuous. For unbounded $G(G_1)$, we assume in addition that the functions F, φ increase not faster than $\exp(a|x|^2)$, as $|x| \to \infty$; here, a is a positive constant, which depends on T.

Let

$$u_0(x,t) = J_1(x,t) + J_2(x,t), \tag{4.71}$$

where

$$J_1(x,t) = \int_0^t dt' \int_{G(G_1)} F(x',t') Z_0(x-x', t-t') dx',$$

$$J_2(x,t) = \int_{G(G_1)} \varphi(x') Z_0(x-x', t) dx'.$$

It is not difficult to verify that u_0 satisfies equation (4.68) and initial condition (4.69). Consequently, the function $u_1 = u - u_0$ solves the following problem:

$$\begin{aligned} \frac{\partial u_1}{\partial t} &= \Delta u_1, \; x \in G(G_1), \; 0 < t < T, \\ u_1(x,0) &= 0, \\ Bu_1|_\Gamma &= g_1(y,t), \quad y \in \Gamma, \end{aligned} \tag{4.72}$$

where

$$g_1(y,t) = g(y,t) - Bu_0(y,t).$$

Thus, if the functions u_0 and $Bu_0|_\Gamma$ are known, a solution to inhomogeneous problem (4.68), (4.70) can be found by solving homogeneous problem (4.72). Denote by $\bar{F}(x,t)$ and $\bar{\varphi}(x)$ the continuations of $F(x,t)$ and $\varphi(x)$, respectively,

$$\bar{F}(x,t) = \begin{cases} F(x,t), & x \in G \\ 0, & x \notin G \end{cases}, \tag{4.73}$$

$$\bar{\varphi}(x,t) = \begin{cases} \varphi(x), & x \in G \\ 0, & x \notin G \end{cases}. \tag{4.74}$$

Hence,

$$J_1(x,t) = \int_0^t dt' \int_{\mathbb{R}^m} \bar{F}(x',t') Z_0(x-x',t-t')dx',$$

$$J_2(x,t) = \int_{\mathbb{R}^m} \bar{\varphi}(x') Z_0(x-x',t-t')dx'. \tag{4.75}$$

It is clear that

$$J_1(x,t) = \mathbb{E}\eta_1^{(0)}(x,t),$$

where

$$\eta_1^{(0)}(x,t) = t\bar{F}(x + 2[(t-\theta)\gamma_{m/2}]^{1/2}\omega, t-\theta). \tag{4.76}$$

Here θ is a random variable uniformly distributed on $[0,t]$, and ω is a unit isotropic vector in $\mathbb{R}^m$.

Analogously,

$$\frac{\partial J_1(x,t)}{\partial x_i} = \mathbb{E}\eta_1^{(i)}(x,t),$$

where

$$\eta_1^{(i)} = -\sqrt{t}\frac{2\Gamma\left(\frac{m+1}{2}\right)}{\Gamma\left(\frac{m}{2}\right)}\bar{F}\left(x + 2\left(t - \frac{\theta^2}{t}\right)^{1/2} \cdot \sqrt{\frac{\gamma_{m+1}}{2}}\omega, \frac{\theta^2}{t}\right)\omega_i,\ i = 1,\dots,m, \tag{4.77}$$

and

$$J_2(x,t) = \mathbb{E}\eta_2^{(0)}(x,t),$$
$$\frac{\partial J_2(x,t)}{\partial x_i}(x,t) = \mathbb{E}\eta_2^{(i)}(x,t), \; i = 1,2,\dots,m,$$

where

$$\eta_2^{(0)}(x,t) = \bar{\varphi}(x + 2\sqrt{t\gamma_{m/2}}\,\omega),$$
$$\eta_2^{(i)} = -\frac{1}{\sqrt{t}}\frac{\Gamma\left(\frac{m+1}{2}\right)}{\Gamma(m/2)}\bar{\varphi}(x + 2\sqrt{t\gamma_{\frac{m+1}{2}}}\,\omega)\omega_i, \; i = 1,2,\dots,m.$$

4.9 Calculation of derivatives on the boundary

We deal here with the problem of calculating the normal derivative $\frac{\partial u}{\partial n}(y)$ for the solution of the following BVP:

$$\begin{aligned}\frac{\partial u}{\partial t}(x,t) &= \Delta u(x,t) + f(x,t), \;\; x \in G, \;\; t > 0,\\ u(x,0) &= \varphi(x), \quad x \in G,\\ u(y,t) &= 0, \quad y \in \Gamma.\end{aligned} \tag{4.78}$$

Note that the straightforward differentiation of the double-layer potential representation, W, leads to the divergent integral as $x \to y \in \Gamma$. To overcome this difficulty, we can use the specific feature of problem (4.78), namely, the condition $u(y,t) = 0$, $y \in \Gamma$. Denote by $G^0(x,t;x')$ the Green's function, which is defined as the solution to the following BVP:

$$\begin{aligned}\frac{\partial G^0}{\partial t}(x,t;x') - \Delta_x G^0(x,t;x') &= \delta(x - x',t), \; x \in G, \; t > 0,\\ G^0(x,0;x') &= 0, \quad x \in G, \quad x \neq x',\\ G^0(y,t;x') &= 0, \quad y \in \Gamma, \quad t > 0.\end{aligned}$$

Then, we have

$$\begin{aligned}u(x,t) = &\int_0^t dt' \int_G f(x',t')G^0(x,t-t';x')dx'\\ &+ \int_G \varphi(x')G^0(x,t;x')dx'.\end{aligned} \tag{4.79}$$

Let

$$G^0(x,t;x') = Z_0(x-x',t) + g(x,t;x'),$$

where g is a continuous in $x \in G$ function.

Then, we can rewrite (4.79) in the following form:

$$\begin{aligned} u(x,t) &= \int_0^t dt' \int_G g(x,t-t';x')f(x',t')dx' \\ &+ \int_G \varphi(x')g(x,t;x')dx' + \int_0^t dt' \int_G Z_0(x-x',t-t')f(x',t')dx' \\ &+ \int_G \varphi(x')Z_0(x-x',t)dx', \end{aligned} \tag{4.80}$$

where the function g has to satisfy the BVP:

$$\begin{aligned} &\frac{\partial g}{\partial t}(x,t;x') = \Delta_x g(x,t;x'), \quad x \in G, \quad t > 0, \\ &g(x,0;x') = 0, \qquad x \in G, \\ &g(y,t;x') = -Z_0(y-x',t), \quad y \in \Gamma, \quad t > 0. \end{aligned} \tag{4.81}$$

We continue $g(x,t;x')$ to the whole space $\mathbb{R}^m$ by setting

$$g(x,t;x') = -Z_0(x-x',t), \qquad (x,t) \in G_1 \times (0,\infty).$$

Since $x' \notin G_1$, the function g satisfies in $G_1 \times (0,\infty)$ the following BVP:

$$\begin{aligned} &\frac{\partial g}{\partial t}(x,t;x') = \Delta_x g(x,t;x'), \quad x \in G_1, \quad t > 0, \\ &g(x,0;x') = 0, \qquad x \in G_1, \\ &\frac{\partial g}{\partial n(y)}(y,t;x') = -\frac{\partial Z_0}{\partial n(y)}(y-x',t) = \psi(y,t;x'), \ y \in \Gamma, \ t > 0. \end{aligned} \tag{4.82}$$

Consequently, the function g can be represented in $G_1 \times (0,\infty)$ as a single-layer potential

$$g(x,t;x') = 2\int_0^t d\tau \int_\Gamma Z_0(x-y,t-\tau)\mu(y,\tau;x')d\sigma(y), \tag{4.83}$$

where the density μ solves in $\Gamma \times (0,\infty)$ the integral equation

$$\begin{aligned} \mu(y,\tau;x') = &-\int_0^\tau d\tau' \int_\Gamma \frac{|y-y'|}{|\tau-\tau'|} \cos\varphi_{yy'} Z_0(y-y',\tau-\tau')\mu(y',\tau';x')d\sigma(y') \\ &+ \psi(y,\tau;x'). \end{aligned} \tag{4.84}$$

The representation for $g(x,t;x')$ is true in $G_1\times(0,\infty)$ and for $(x,t)\in G\times(0,\infty)$ as well, since potential (4.83) satisfies the heat equation and the boundary condition in (4.82). Hence, we get

$$\begin{aligned} u(x,t) &= \int_0^{\tau} dt' \int_G f(x',t')\left[2\int_0^{t-t'} d\tau \int_\Gamma Z_0(x-y,t-t'-\tau)\mu(y,\tau;x')d\sigma(y)\right] dx' \\ &+ \int_G \varphi(x')\left[2\int_0^{t} d\tau \int_\Gamma Z_0(x-y,t-\tau)\mu(y,\tau;x')d\sigma(y)\right] dx' \\ &+ J_1(x,t) + J_2(x,t). \end{aligned} \tag{4.85}$$

Here

$$\begin{aligned} J_1(x,t) &= \int_0^{t} dt' \int_G Z_0(x-x',t-t')f(x',t')dx', \\ J_2(x,t) &= \int_G \varphi(x')Z_0(x-x',t)dx'. \end{aligned}$$

Calculate now the normal derivative of the solution to (4.78) by differentiating at $x\in G$ in (4.79) and taking the limit as $x\to y\in\Gamma$:

$$\begin{aligned} \frac{\partial u}{\partial n(y)}(y,t) &= \lim_{x\to y}\sum_{i=1}^{m}\frac{\partial u}{\partial x_i}(x,t)n_i(y) \\ &= \int_0^{t} dt' \int_G f(x',t') \\ &\times\left[\mu(y,t-t';x') + \int_0^{t-t'} d\tau \int_\Gamma 2\frac{\partial Z_0}{\partial n(y)}(y-y',t-t'-\tau)\mu(y',\tau;x')d\sigma(y')\right] dx' \\ &+ \int_G \varphi(x')\left[\mu(y,t;x') + \int_0^{t} d\tau \int_\Gamma 2\frac{\partial Z_0}{\partial n(y)}(y-y',t-\tau)\mu(y',\tau;x')d\sigma(y')\right] dx' \\ &+ \frac{\partial J_1}{\partial n(y)} + \frac{\partial J_2}{\partial n(y)}. \end{aligned} \tag{4.86}$$

Here, we have used the limit properties of the normal derivative of the single-layer potential (Section 4.1).

Thus, the normal derivative is represented as a linear functional of the solution to (4.84). Let us suppose for simplicity that the domain is strongly convex (i.e., the curvature is positive and separated from zero) and $\varphi(x)\equiv 0,\ x\in G$. Monte Carlo estimators for

$$\int_0^t dt' \int_G f(x',t')\mu(y,t-t';x')dx', \quad \frac{\partial J_1}{\partial n} \quad \text{and} \quad \frac{\partial J_2}{\partial n}$$

are already constructed in Section 4.2. Therefore, we need to describe the algorithm for calculating

$$J_3(y,t) = \int_0^t dt' \int_G f(x',t') \left[\int_0^{t-t'} d\tau \int_\Gamma 2\frac{\partial Z_0}{\partial n(y)}(y-y',t-t'-\tau)\cdot\mu(y',\tau;x')d\sigma(y') \right] dx'.$$

Let $p(x',t')$ be a density in $G\times(0,t)$ consistent with $f(x',t')$. Then, we can use the following random estimator:

$$\eta_{yt} = \sum_{n=0}^{N_{t-t'}} (-1)^n Q_n \psi(y_n,\tau_n;x'), \tag{4.87}$$

where $\{(y_n,\tau_n)\}_{n=0}^{\infty}$ is F-process with the initial density

$$\begin{aligned} p_0(y_0,\tau_0) &= 2\frac{\partial Z_0}{\partial n(y_0)}(y_0-y,t-t'-\tau_0) \\ &= \frac{|y-y_0|}{|t-t'-\tau_0|}\cos\varphi_{y_0y} Z_0(y_0-y,t-t'-\tau_0), \end{aligned}$$

and the random weights Q_n are calculated from

$$Q_0 = \frac{\cos\varphi_{yy_0}}{\cos\varphi_{y_0y}}, \quad Q_n = Q_{n-1}\frac{\cos\varphi_{y_{n-1}y_n}}{\cos\varphi_{y_ny_{n-1}}},$$

$n = 1,2,\ldots,N_{t-t'} = \max\{n : \tau_n > 0\}$.

Denote by ω the random trajectory $\{(y_n,\tau_n)\}_{n=0}^{N_{t-t'}}$ and by $\mathbb{E}_\omega$ the expectation over its distribution. The random variable η_{yt} depends on the parameters x', t' and ω. By the construction, we get

$$\mathbb{E}_\omega\{\eta_{yt}|(x',t')\} = \int_0^{t-t'} d\tau \int_\Gamma \frac{\partial Z_0}{\partial n(y)}(y-y',t-t'-\tau)\mu(y',\tau;x')d\sigma(y').$$

Consequently,

$$\zeta_{yt} = \frac{\eta_{yt} f(x',t')}{p(x',t')}$$

is an unbiased estimator for $J_3(y,t)$:

$$\mathbb{E}\zeta_{yt} = J_3(y,t).$$

Next, we analyse the variance. By definition,

$$\mathbb{E}\zeta_{yt}^2 = \mathbb{E}_{(x',t')}\left[\frac{f^2(x',t')}{p^2(x',t')}\,\mathbb{E}_\omega\,\{\eta_{yt}^2|(x',t')\}\right]. \tag{4.88}$$

Lemma 4.17. *Suppose that the function f is chosen so that*

$$f(x,t) \equiv 0, \quad x \in \Gamma_\delta,$$

where

$$\Gamma_\delta = \{x : x \in G; \inf_{y\in\Gamma} |x-y| < \delta\}$$

is the δ-neighbourhood of Lyapunov continuous Γ (δ-strip near the boundary). Choose the density $p(x',t')$ in such a way that

$$\int_0^t dt' \int_G \frac{f^2(x',t')}{p(x',y')}\,dx' < \infty.$$

Then, the variance of ζ_{yt} is finite.

Proof. Indeed, from (4.84), we get

$$\sup_{x'\in G\setminus\Gamma_\delta} |\mu(y,\tau;x')| < c,$$

where c is a constant that depends on $\sup|\psi(y,\tau;x')|$, on the form of G and on δ. From this, we get the uniform estimation:

$$\mathbb{E}\{\eta_{yt}^2|(x',t')\} < C_1, \quad (x',t') \in (G\setminus\Gamma_\delta)\times(0,t).$$

Hence,

$$\mathbb{E}\zeta_{yt}^2 = \int_0^t dt' \int_{G\setminus\Gamma_\delta} \frac{f^2(x',t')}{p(x',t')}\,\mathbb{E}\,\{\eta_{yt}^2|(x',t')\}dx' < \infty.$$

□

5 Spatial problems of elasticity

5.1 Elastopotentials and systems of boundary integral equations of the elasticity theory

In this chapter we consider differential equations of elasticity theory and use theoretical results given in [58]. Let $u(x)$ be a vector function of displacements, $u : G \to \mathbb{R}^3$, $u \in C^2(G) \cap C(\overline{G})$, where G is a bounded simple connected domain in $\mathbb{R}^3$. It is well known then that a vector of displacements satisfies the Lamé equations

$$\Delta^* u(x) \equiv \mu \Delta u(x) + (\lambda + \mu) \operatorname{grad} \operatorname{div} u(x) = 0, \quad x \in G, \tag{5.1}$$

where λ, μ are positive constants that characterize an elastic matter in the domain G. Commonly, these coefficients are called the Lamé parameters.

Let $\Gamma = \partial G$ be a Lyapunov simple connected surface, and $\varphi(y)$, $y \in \Gamma$ is some continuous vector function on the boundary. Then, we can consider the first BVP for the Lamé equations

$$u(y) = \varphi(y), \quad y \in \Gamma, \tag{5.2}$$

and the second BVP

$$Tu(y) = \varphi^*(y), \quad y \in \Gamma, \tag{5.3}$$

where $T = \{T_{ij}\}$, $i, j = 1, 2, 3$, is the tension operator of the classical elasticity theory,

$$T_{ij} = \lambda n_i(y) \frac{\partial}{\partial y_j} + \mu n_j(y) \frac{\partial}{\partial y_i} + \mu \delta_{ij} \frac{\partial}{\partial n(y)}.$$

Here, δ_{ij} is the Kronecker symbol, and $n(y) = (n_1(y), n_2(y), n_3(y))^T$ is, as usual, the **interior** normal vector at a point $y \in \Gamma$.

Vector equation (5.1) is an elliptic system with constant coefficients, and its fundamental solutions can be written out explicitly. One of the classical fundamental solutions is the Kelvin matrix

$$S[x, y] = \{s_{ij}(x, y)\}, \; i, j = 1, 2, 3,$$

where

$$s_{ij}(x, y) = \lambda' \frac{\delta_{ij}}{|x - y|} + \mu' \frac{(x_i - y_i)(x_j - y_j)}{|x - y|^3},$$
$$\lambda' = (\lambda + 3\mu)(4\pi\mu(\lambda + 2\mu))^{-1},$$
$$\mu' = (\lambda + \mu)(4\pi\mu(\lambda + 2\mu))^{-1}.$$

It means that

$$\Delta_x^{\star} S = \delta(x-y)E,$$

where $E = \{\delta_{ij}\}$ is the unit matrix.

In the full analogy with a scalar case of the Laplace equation, we can introduce vector surface potentials (elastopotentials): a single-layer potential,

$$V^{\star}(x) = \int_{\Gamma} S[x,y]\Psi(y)d\sigma(y), \tag{5.4}$$

and a double-layer potential,

$$W^{\star}(x) = \int_{\Gamma} K[x,y]\Phi(y)d\sigma(y). \tag{5.5}$$

Here $K[x,y] = \{k_{ij}(x,y)\}$, $i,j = 1,2,3$,

$$\begin{aligned} k_{ij}(x,y) &= \frac{\delta_{ij}}{2\pi} \cdot \frac{\partial}{\partial n(y)} \frac{1}{|x-y|} \\ &\quad + \mu \sum_{k=1}^{3} M_{jk}\left((\lambda' - \mu') \frac{\delta_{ki}}{|x-y|} + 2\mu \frac{(x_k - y_k)(x_i - y_i)}{|x-y|^3} \right), \\ M_{jk} &= n_k(y)\frac{\partial}{\partial y_j} - n_j(y)\frac{\partial}{\partial y_k}, \quad k,j = 1,2,3. \end{aligned}$$

Performing all differentiations and calculations, we can write out k_{ij} in the explicit form

$$\begin{aligned} k_{ii}(x,y) &= \frac{1}{2\pi} \cdot \frac{1}{\lambda + 2\mu} \cdot \frac{\cos\varphi_{yx}}{|x-y|^2} \left(\mu + 3(\lambda+\mu) \frac{(x_i - y_i)^2}{|x-y|^2} \right), \\ k_{ij}(x,y) &= \frac{3}{2\pi} \cdot \frac{\lambda+\mu}{\lambda+2\mu} \cdot \cos\varphi_{yx} \cdot \frac{(x_i - y_i)(x_j - y_j)}{|x-y|^4} \\ &\quad + \frac{\mu}{\pi(\lambda+2\mu)} \cdot \frac{n_i(y)(x_j - y_j) - n_j(y)(x_i - y_i)}{|x-y|^3}, \quad i \neq j. \end{aligned} \tag{5.6}$$

Denote by $k_{ij}^{(r)}$ the first term in the last sum, and by $k_{ij}^{(s)}$, we denote the second term. The notation is related to different properties of these kernels when $|x-y| \to 0$, $x, y \in \Gamma$. On Lyapunov surfaces, we have

$$\begin{aligned} k_{ij}^{(r)}(x,y) &= 0\left(|x-y|^{-2+\gamma}\right), \\ k_{ij}^{(s)}(x,y) &= 0\left(|x-y|^{-2}\right), \end{aligned}$$

where γ is the Lyapunov constant. Hence, $k_{ij}^{(r)}$ is a weakly polar kernel and $k_{ij}^{(s)}$ is a singular kernel.

Theorem 5.1. [58] *If densities Φ, Ψ are Hölder continuous vector functions on Γ with exponent equal to β : $\Phi, \Psi \in C^{0,\beta}(\Gamma)$, then*
(1) *elastopotentials V^*, W^* satisfy the Lamé equation*

$$\Delta^* V^*(x) = \Delta^* W^*(x) = 0$$

at every point $x \notin \Gamma$;
(2) *the limiting values of W^* satisfy the relation*

$$(W^*(x))_\pm = \mp\Phi(x) + W^*(x), \quad x \in \Gamma, \tag{5.7}$$

where $W^*(x)$ is the direct value of a double-layer potential;
(3) *the limiting values of the tension operator applied to V^* satisfy*

$$(TV^*(x))_\pm = \pm\Psi(x) + TV^*(x), \quad x \in \Gamma. \tag{5.8}$$

Following this theorem, we seek a solution to problem (5.1), (5.2) in the form of a double-layer elastopotential $W^*(x)$. This leads to the integral equation of the second kind for an unknown vector density, Φ:

$$\Phi(y) = -\kappa \int_\Gamma K[y, y']\Phi(y')d\sigma(y') + \varphi(y), \quad \kappa = 1, \tag{5.9}$$

or in the operator form

$$\Phi = -\kappa\mathcal{K}\Phi + \varphi.$$

Consider second BVP (5.3) for Lamé equations (5.1) in G_1 and seek its solution in the form of a single-layer elastopotential $V^*(x)$. Then, in the full analogy with the classical Neumann problem for the Laplace equation, we come to the integral equation of the second kind for the vector density Ψ of this potential

$$\Psi(y) = -\kappa \int_\Gamma K[y', y]\Psi(y')d\sigma(y') + \varphi^*(y), \tag{5.10}$$

which is adjoint to (5.9).

Considering $\mathcal{K}$ and $\mathcal{K}^*$ as integral operators in the space of vector functions $C^{0,\beta}(\Gamma)$, we can assert that they, despite their singularity, have the same spectral properties as the corresponding integral operators (3.6) and (3.22) of the classical potential theory. Hence, the Neumann series for operators $\mathcal{K}, \mathcal{K}^*$ do not converge, but equations (5.9), (5.10) have the unique solutions. These solutions can be found with the help of the same transformations and analytical continuation methods that have been used in Chapter 3 in the case of the BVPs for the Laplace equation. As a consequence, we can write out a solution of the first interior problem for the Lamé equations in the form of a finite sum

$$u(x) = \sum_{i=0}^{n} l_i^{(n)} \int_{\Gamma} K[x,y]\mathcal{K}^i \varphi(y) d\sigma(y) + \delta^{(n)}(x), \tag{5.11}$$

where $l_i^{(n)}$ are determined by the analytical continuation method (Chapters 2 and 3). It is essential to note that in this case, a proof of the convergence of the transformed series is significantly more complex problem since analyticity of the matrix resolvent for singular integral equation needs its special and careful investigation [58].

5.2 First BVP and estimators for singular integrals

Considering first BVP (5.2) for the Lamé equations (5.1), we seek its solution in the form of a double-layer elastopotential W^*. As a consequence, we come to the need of computing the iterations for the boundary matrix-integral singular operator, $\mathcal{K}$.

It is obvious that every unbiased estimator ξ_i or ξ_i^* (Chapter 2) has infinite moments $\mathbb{E}|\xi_i|^m$, $m \geq 1$ in the case when a kernel of an integral operator is singular. Further in this chapter, we construct an unbiased estimator for a singular part of the integral operator $\mathcal{K}$ with finite variance. To do that, we will use the kernel symmetry properties.

Consider a singular integral operator acting on functions defined on a Lyapunov surface $\Gamma \in \mathbb{R}^3$,

$$K^{(s)}\varphi(y) = \int_{\Gamma} k(y,y')\varphi(y')d\sigma(y'). \tag{5.12}$$

Here, singularity means that $|k(y,y')| = O(|y-y'|^{-2})$ for $|y-y'| \to 0$. So this integral does not exist in the ordinary sense. By definition, we have

$$K^{(s)}\varphi(y) = \lim_{\varepsilon \to 0} \int_{\Gamma^{(\varepsilon)}} k(y,y')\varphi(y')d\sigma(y'),$$

if the limit in the right-hand side exists. Here,

$$\Gamma^{(\varepsilon)} = \Gamma \setminus \{y' : |y-y'| \leq \varepsilon\}.$$

It is well known [70] that both necessary and sufficient condition of existence for $K^{(s)}\varphi(y)$ is the following equality:

$$\int_0^{2\pi} k_0(y,\psi)d\psi = 0, \tag{5.13}$$

where

$$k_0(y,\psi) = \lim_{y'\to y} k(y,y') \cdot |y-y'|^2,$$

and y' approaches y in such a way that

$$\psi = \angle(\pi(y') - y, e) \tag{5.14}$$

remains constant. Here, we denote by $\pi(y')$ the projection of y' on the tangent plane at the point y, and e is some fixed direction on this plane. It is obvious that to construct an effective estimator, we have to use condition (5.13). One of the possible ways of its utilizing is the method proposed in the study of Wagner [121]. This algorithm is a variant of a stratified sampling method [23], and it can be applied in the case when we exactly know the region $\Gamma_+(y)$ where the kernel $k(y,y')$ is positive, and the corresponding region $\Gamma_-(y)$ where this function is negative.

Different techniques will be employed here. It is based on some symmetrical mappings, which are used for constructing walk-on-boundary branching Markov chain. To define these mappings, we introduce the local spherical coordinates with the origin at the point y in such a way that $y' = (r,\theta,\psi)$ where $r = |y-y'|$, ψ is defined in (5.14) and θ is the angle between the vector $y'-y$ and the tangent plane at the point y. So, when $y' \in \Gamma$, then $\alpha_y(y')$ is defined as such a point on Γ that

$$|y - \alpha_y(y')| = r,$$

and

$$\angle(\pi(\alpha_y(y') - y), e) = \psi + \pi.$$

It means that $\alpha_y(y') = (r, \bar\theta, \psi + \pi)$, where $\bar\theta = \theta(1 + 0(r^\gamma))$, as a consequence of the Lyapunov's continuity of the boundary.

Introduce another symmetrical mapping. Let us define

$$\chi_y(y') = (\bar r, \theta, \psi + \pi).$$

It means that the angle between the vector $\chi_y(y') - y$ and the tangent plane is equal to θ and $\bar r$ is obtained from the condition that $\chi_y(y')$ lies on Γ. Obviously, $\bar r = r(1 + 0(r^\gamma))$ for a Lyapunov boundary.

Consider now regular and singular parts of the kernel $k_{ij}(y,\cdot)$ where the second argument is obtained as a result of a symmetrical mapping applied to y'. Without loss of generality, the point y may be considered as either elliptic or an interior point of a plane part of the boundary. In the first case,

$$\begin{aligned} k_{ij}^{(r)}(y, \alpha_y(y')) &= k_{ij}^{(r)}(y,y')(1 + 0(r^\gamma)),\\ k_{ij}^{(s)}(y, \alpha_y(y')) &= -k_{ij}^{(s)}(y,y')(1 + 0(r^\gamma)), \end{aligned} \tag{5.15}$$

and hence

$$k_{ij}(y, \alpha_y(y')) = -k_{ij}(y, y')(1 + O(r^{\gamma})).$$

The same relations are valid for $\chi_y(y')$ also.

It is obvious enough that points $\alpha_y(y')$ and $\chi_y(y')$ coincide in the case when y, y' lie in the same plane and symmetrical mapping does not move us out of it. We have $k^{(r)}(y, y') = k^{(r)}(y, \alpha_y(y')) = 0$ here, since $n(y')$ is orthogonal to both $y - y'$ and $y - \alpha_y(y')$. Hence,

$$k_{ij}(y, \alpha_y(y')) = k_{ij}^{(s)}(y, \alpha_y(y')) = -k_{ij}(y, y'). \tag{5.16}$$

Fix now some point $y \in \Gamma$ and construct an estimator for $\mathcal{K}\varphi(y)$ where φ is a Hölder continuous function on Γ. To do that, we have to choose some distribution density for a point y'.

At first, we consider y to be an elliptic point. In this case, it is more convenient to define the required distribution density not in the surface measure σ but in the angle measure, Ω:

$$d\Omega(y') = \frac{\cos\varphi_{y'y}}{|y - y'|^2} d\sigma(y').$$

Direction $y' - y$ is determined by two angles, θ and ψ. ψ is chosen randomly (uniformly) on the segment $(0, 2\pi)$ and θ is sampled proportionally to $(\sin\theta)^{-\beta}$ for some $\beta \geq 0$. So we have

$$\bar{p}^{(\beta)}(\theta, \psi) = \text{const} \cdot (\sin\theta)^{\beta}.$$

Note that if $\beta = 0$, then y' is isotropically distributed in the solid angle with its vertex at the point y. This means that for $\beta = 0$, the distribution density of y' in the surface measure,

$$p^{(\beta)}(y, y') = \bar{p}^{(\beta)} \cdot \frac{\cos\varphi_{y'y}}{|y - y'|^2}, \tag{5.17}$$

is equal to the transitional density (3.10) of an isotropic WB-process. Hence, we have

$$p^{(\beta)}(y, y') = O(|y - y'|^{-2+\gamma-\beta}), \tag{5.18}$$

when $|y - y'| \to 0$. Thus, we have to choose $\beta < \gamma$. Recall that we used such non-isotropic densities in the third and fourth chapters.

Suppose now that y is located in the plane part of the boundary. It means that there exists such $r_0 > 0$ that the circle

$$C = \{y' : y' \in \Gamma, |y - y'| < r_0\} \tag{5.19}$$

is totally contained in the same plane. We define the distribution density in this circle by the formula

$$p^{(\beta)}(y, y') = \text{const} \cdot |y - y'|^{-1-\beta}, \tag{5.20}$$

for some $\beta \geq 0$. It means that angle ψ is sampled isotropically, and constant factor is chosen in such a way that

$$\int_C p^{(\beta)}(y, y') d\sigma(y') = 1.$$

Next, we define a transitional density $p(y, y')$ by the following procedure. Let ω be a positive number less than 1. With probability ω, we sample y' in the neighbourhood C of the point y (see (5.19)) with the density $p^{(\beta)}$, defined in (5.17) or (5.20), depending on the geometrical properties of y. With the complementary probability, $1 - \omega$, point y' is chosen randomly in $\Gamma \setminus C$ in accordance with some convenient density p_1, which may, for example, be equal to the transitional density of an isotropic WB-process. So we have

$$p(y, y') = \omega p^{(\beta)}(y, y') \cdot I_C(y') + (1 - \omega) p_1(y, y') \cdot I_{\Gamma \setminus C}(y'), \tag{5.21}$$

where I_C is the indicator function of set C.

Recall now that our goal is to construct an estimator for a singular integral, and this is the purpose of introducing the density (5.21). Denote $y_0 = y$, $y_{11} = y'$ and $y_{12} = \alpha_y(y')$ (or $\chi_y(y')$) and consider the estimator

$$\zeta_1^*(y_0) = -\frac{1}{2\omega} \left[\frac{K[y_0, y_{11}]\varphi(y_{11})}{p^{(\beta)}(y_0, y_{11})} + \frac{K[y_0, y_{12}]\varphi(y_{12})}{p'(y_0, y_{12})} \right], \tag{5.22}$$

if $y' \in C$, and

$$\zeta_1^*(y_0) = -\frac{1}{1 - \omega} \cdot \frac{K[y_0, y_{11}]\varphi(y_{11})}{p_1(y_0, y_{11})}, \tag{5.23}$$

for $y' \in \Gamma \setminus C$. It means that we use the transitional density (5.21) and bifurcate the trajectory of WB-process $\{y_i\}$ in the case when y_{i+1} is located in the neighbourhood of the point y_i. In (5.22), we denote by $p'(y_0, y_{12})$ the distribution density of the point y_{12}. This point is obtained as the result of some symmetrical mapping applied to y_{11} and hence, in the general case, its distribution depends on the geometrical properties of Γ in C. If C, for example, is a part of a sphere or of a plane, then $p'(y_0, y_{12}) = p^{(\beta)}(y_0, y_{11})$ for both α and χ. By definition, angle ψ (one of the angle coordinates of the point y_{11}) is isotropically distributed, and hence $\psi + \pi$ is also isotropically distributed. As a consequence, angle distributions of the points y_{11} and $\chi_y(y_{11})$ coincide. So we have in this case,

$$p'(y_0, y_{12}) = p^{\beta}(y_0, y_{11}) \cdot \frac{\cos \varphi_{y_{12} y_0}}{\cos \varphi_{y_{11} y_0}} \cdot \frac{r^2}{\bar{r}^2}.$$

As for the distribution p', which is considered in the surface measure, σ, we can assert only that $p'(y_0, y_{12}) = p^{(\beta)}(y_0, y_{11}) \cdot (1 + O(r^{\gamma}))$, where γ is the boundary Lyapunov constant.

It is obvious that

$$\mathbb{E}\zeta_1^{\star}(y_0) = \mathcal{K}\varphi(y_0).$$

Next, we establish that the variance of this estimator is finite. To do that, it is sufficient to consider a singular part of a non-diagonal element of matrix $K[y_0, y_{11}]$ and show that an estimator

$$\xi(y_0) = \frac{1}{2}\left[\frac{k_{ij}^{(s)}(y_0, y_{11})\varphi_j(y_{11})}{p^{(\beta)}(y_0, y_{11})} + \frac{k_{ij}^{(s)}(y_0, y_{12})\varphi_j(y_{12})}{p'(y_0, y_{12})}\right]$$

for the integral, $\int_{\Gamma} k_{ij}^{(s)}(y_0, y_1)\varphi_j(y_1)d\sigma(y_1)$, has finite second moment. Suppose that $\varphi_j \in C^{0,s}(\Gamma)$, i.e., Hölder continuous with the exponent s. Consider the function $\zeta = \xi^2(y_0)p^{(\beta)}(y_0, y_{11})$ when $r = |y_0 - y_{11}|$ tends to zero. Applying (5.15) and then (5.18), we come to the following relation in the case when y is an elliptic point:

$$\zeta = \frac{1}{4} \cdot \frac{(k_{ij}^{(s)}(y_0, y_{11}))^2}{p^{(\beta)}(y_0, y_{11})} \cdot \varphi_j^2(y_{11}) \cdot O(r^{2\min(\gamma,s)}) = O(r^{-2-\gamma+\beta+2\min(\gamma,s)}),$$

while for the case when y is located in a plane part of Γ, we have

$$\zeta = \frac{1}{4} \cdot \frac{(k_{ij}^{(s)}(y_0, y_{11}))^2}{p^{(\beta)}(y_0, y_{11})} \cdot \varphi_j^2(y_{11}) \cdot O(r^{2s}) = O(r^{-3+\beta+2s}).$$

Thus, to guarantee the variance of a constructed estimator is finite, the choice of β should be adjusted to the smoothness properties of both Γ and the function under the integral sign. In the elliptic case, we must have $\gamma > \beta > \gamma - 2\min(s, \gamma)$, and in the plane case, the following condition should be satisfied: $1 > \beta > 1 - 2s$. As a consequence, the isotropic transitional density can be used for an elliptic point y if $s \geq \gamma/2 + \varepsilon$ for some $\varepsilon > 0$. In the case when Γ is strictly convex, the isotropic density on the whole boundary can be used.

So we see that if we construct the estimator in accordance with formulas (5.22), (5.23) and use the transitional density (5.21) for simulation of WB-process, then on every step, we have

$$\mathbb{E}(\zeta_i^{\star}(y_0)/y_0) = \mathcal{K}\zeta_{i-1}^{\star}(y_0),$$

and moreover, the conditional second moment of $\zeta_i^\star$ is finite. It means that the estimator $\zeta_i^\star$ is constructed recurrently according to the following formulas:

$$\zeta_i^\star(y_0) = -\frac{1}{2\omega}\left[\frac{K[y_0,y_{11}]\zeta_{i-1}^\star(y_{11})}{p^{(\beta)}(y_0,y_{11})} + \frac{K[y_0,y_{12}]\zeta_{i-1}^\star(y_{12})}{p'(y_0,y_{12})}\right],$$

for $y' = y_{11}$ in the neighbourhood C of the point y_0, and

$$\zeta_i^\star(y_0) = -\frac{1}{1-\omega}\cdot\frac{K[y_0,y_{11}]\zeta_{i-1}^\star(y_{11})}{p_1(y_0,y_{11})},$$

for $y' \in \Gamma \setminus C$.

The WB-process starts from the point on Γ, y_0, which can be sampled using the initial distribution density p_0 from (5.9), that is, isotropically in the solid angle with its vertex at the point $x_0 \in G$. We can take $x_0 = x$, the point, in which we want to calculate the solution of BVP (5.1), (5.2). Next, we use the transitional density (5.21) and construct a Markov chain, branching on every step with some probability, ω. Then, we have that

$$\xi_i^\star = \frac{K[x,y_0]}{p_0(y_0)}\zeta_i^\star(y_0)$$

is an unbiased estimator for the iteration of the integral operator, $\mathcal{K}$:

$$\mathbb{E}\xi_i^\star = (K, \mathcal{K}^i\varphi),$$

and its variance is finite.

Substituting $\xi_i^\star$ in (5.11), we obtain a $\delta^{(n)}$-biased estimator for the solution to the BVP, $u(x)$:

$$\xi^\star = \sum_{i=0}^{n} l_i^{(n)}\xi_i^\star,$$

where $\delta^{(n)}(x)$ is a vector function whose norm is decreasing as $n \to \infty$ at the speed of a geometric progression.

5.3 Other BVPs for the Lamé equations and regular integral equations

Consider the second BVP (5.3) for the Lamé equations in the exterior $G_1 \subset \mathbb{R}^3$ of a bounded domain G with a simple connected boundary. We seek its solution in the form of a single-layer elastopotential $V^\star$. As a consequence, we come to the integral equation (5.10) for an unknown potential density Ψ. This integral equation is adjoint to (5.9). So in the full analogy with a scalar case, we can use the same WB-process that has been simulated to obtain an estimator for the first BVP and construct a direct vector estimator. From here we have

$$\mathbb{E}\xi_i = (\mathcal{K}^{*i}\varphi^*, S),$$

and

$$\xi = \sum_{i=0}^{n} l_i^{(n)} \xi_i$$

is a $\delta^{(n)}$-biased estimator for $u(x)$ where

$$\xi_i = \zeta_i[x, y_0]\frac{\varphi^*(y_0)}{p_0(y_0)},$$
$$\zeta_0[x, y_0] \equiv S[x, y_0]$$

and the matrix-valued random weights ζ_i are obtained from the recurrent relation

$$\zeta_i[x, y_0] = -\frac{1}{2\omega}\left[\frac{\zeta_{i-1}[x, y_{11}]K[y_0, y_{11}]}{p^{(\beta)}(y_0, y_{11})} + \frac{\zeta_{i-1}[x, y_{12}]K[y_0, y_{12}]}{p'(y_0, y_{12})}\right],$$

if the Markov chain is branching at the point y_0, and

$$\zeta_i[x, y_0] = -\frac{1}{1-\omega}\cdot\frac{\zeta_{i-1}[x, y_{11}]K[y_0, y_{11}]}{p_1(y_0, y_{11})},$$

otherwise.

Next, in analogy with the Laplace equation, consider the first exterior and the second interior BVPs for the Lamé equations. If we try to seek their solutions in the form of a double-layer elastopotential W^* and a single-layer elastopotential V^*, respectively, we come to integral equations (5.9) and (5.10), where κ is taken to be equal to -1. Since $\kappa = -1$ is a characteristic value of the operator $\mathcal{K}$, and Fredholm theorems are valid for this operator, then it means that solution of integral equations (5.9), (5.10) for $\kappa = -1$ exist if and only if their free terms are orthogonal to all corresponding eigenfunctions of $\mathcal{K}^*$ and $\mathcal{K}$, respectively. It is known [58] that there are six of them, for both of integral operators. So these orthogonality conditions cannot be effectively utilized in the process of an algorithm construction. The conclusion is that if we want to use a WB-process for solving the first exterior and the second interior BVP, then we have to consider other representations of a solution in the form of surface potentials. The first exterior problem, for example, can be solved with a help of a WB-process, if we seek its solution (by analogy with the Laplace equation) as a weighted sum of W^* and V^*.

It is essential to note that in addition to the classical fundamental solution S, there exist other fundamental solutions of the Lamé equations, and as a consequence, different forms of surface potentials can be used. For example, the surface potential

$$W_1^\star(x) = \int_\Gamma K_1[x, y]\Phi_1(y)d\sigma(y)$$

satisfies the Lamé equation at every point $x \notin \Gamma$ and has the same limiting properties when x approaches the boundary. Here, the matrix-valued kernel $K_1[x, y] = \{k_{ij}^{(1)}(x, y)\}$, $i, j = 1, 2, 3$ is defined by the following relations:

$$\begin{aligned} k_{ii}^{(1)} &= \frac{1}{2\pi} \cdot \frac{1}{\lambda + 3\mu} \cdot \frac{\cos\varphi_{yx}}{|x-y|^2} \cdot \left(2\mu - 3(\lambda+\mu)\frac{x_i - y_i}{|x-y|^2}\right), \\ k_{ij}^{(1)} &= \frac{3}{2\pi} \cdot \frac{\lambda+\mu}{\lambda + 3\mu} \cdot \cos\varphi_{yx} \cdot \frac{(x_i - y_i)(x_j - y_j)}{|x-y|^4}, \ i \neq j. \end{aligned} \tag{5.24}$$

Consider the first interior BVP for the Lamé equations and seek its solution in the form of this potential, $W_1^\star$. As the result, we come to the boundary integral equation of the second kind for an unknown vector function Φ_1:

$$\begin{aligned} \Phi_1(y) &= -\kappa \int_\Gamma K_1[y, y']\Phi_1(y')d\sigma(y') + \varphi(y), \\ \kappa &= 1, \ y \in \Gamma. \end{aligned}$$

It is known that this integral equation has the same spectral properties as (5.9), but the kernel K_1 (as we see from (5.21)) does not have singular components. So we can use the same isotropic WB-process that has been constructed in Chapter 3 for solving the Laplace equation. Let $\{y_i, i = 0, 1, 2, \ldots\}$ be a trajectory of this Markov chain, p_0 its initial and p its transitional densities. Then, we can use a standard adjoint vector estimator for the solution $u(x)$:

$$\xi^\star(x; n) = \sum_{i=0}^{n} l_i^{(n)} Q_i^\star \varphi(y_i),$$

where the matrix-valued weights are calculated using the following formulas:

$$Q_0^\star = \frac{K_1[x, y_0]}{p_0(y_0)}, \quad Q_i^\star = -\frac{Q_{i-1}^\star K_1[y_{i-1}, y_i]}{p(y_{i-1}, y_i)},$$

and the coefficients $l_i^{(n)}$ are defined by a method of analytical continuation.

In conclusion to this chapter, we note that there exist many other elasticity problems, which can be treated in an analogous way. This means that some sort of WB-process and, hence, Monte Carlo estimators can be constructed for their solutions.

6 Variants of the random walk on boundary for solving stationary potential problems

6.1 The Robin problem and the ergodic theorem

In all random walk algorithms described in the book up to this point, the solution was represented in the form of an average over a set of independent random trajectories. It is well known, however, that there exists another approach to constructing Monte Carlo algorithms. The methods that exploit such approach date back to the work of Metropolis [67] and are based on a representation of a solution as an average over one long trajectory. Currently, Markov chain Monte Carlo computational methods that stem from this original idea are widely used in various applications (see, e.g. [5, 92]). Further in this section, we describe how this approach, first suggested in [97], can be used for solving the Robin problem.

First, consider the classical problem of electrostatics. Usually, it is formulated the following way: to find the electrostatic potential satisfying the Laplace equation $\Delta\varphi = 0$ outside a system of isolated conductors G_i, such that the natural zero conditions at the infinity are satisfied, $\varphi(r) \to 0$ as $r \to \infty$, and the potential values, $\varphi|_{\Gamma_i} = \varphi_i$, are given on the boundaries of these conductors $\Gamma_i = \partial G_i$.

In applications, the following condition should be taken into account:

$$\int_{\Gamma_i} \frac{\partial\varphi}{\partial n} d\sigma = -4\pi l_i,$$

where l_i is the total charge of the ith conductor. If only one conductor G_1 is given, then the solution can be represented as

$$\varphi = \varphi_1 V(x, y, z),$$

where V is the solution to the first exterior BVP such that $V|_{\partial G_1 = \Gamma_1} = 1$, and φ_1 is determined from

$$\int_{\Gamma_1} \frac{\partial\varphi}{\partial n} d\sigma = -4\pi l_1,$$

i.e., $\varphi_1 = l_1/C$, where

$$C = -\frac{1}{4\pi} \int_{\Gamma_1} \frac{\partial V}{\partial n} d\sigma$$

is the capacitance of the conductor.

For a system of conductors,

$$l_i = C_{ii}\varphi_i + \sum_{j \neq i} C_{ij}(\varphi_j - \varphi_i), \quad C_{ij} = C_{ji}, i = 1, \dots, n,$$

where l_i is the charge, and φ_i is the potential of the ith conductor.

Thus,

$$C_{ij} = l_i^{(j)} = -\frac{1}{4\pi} \int_{\Gamma_i} \frac{\partial \varphi^{(j)}}{\partial n} d\sigma, \quad i, j = 1, \dots, n,$$

where $\varphi^{(j)}$ are harmonic functions that satisfy the following boundary conditions:

$$\varphi^{(j)}|_{\Gamma_i} = 0 \quad (i \neq j), \quad \varphi^{(i)}|_{\Gamma_i} = 1, \quad \varphi^{(j)}|_{\infty} = 0.$$

The problems with multiply connected geometry will be considered in Section 6.3. Here, we restrict ourselves to a simply connected domain $G \equiv G_1$ and construct an estimate based on ergodic properties of a simulated walk-on-boundary Markov chain.

We consider $G \in \mathbb{R}^3$ to be a compact set representing an electrical conductor. For smooth boundary, $\Gamma \equiv \partial G$, there exists the unique solution of the exterior Dirichlet problem we have formulated. It is sufficient, for example, to demand Γ to be a regular piecewise Lyapunov surface [34]. In this case, it is possible to represent V as a single-layer potential

$$V(x) = \int_{\Gamma} \frac{1}{|x - y'|} \mu(y') d\sigma(y'),$$

with a charge density, μ, which is the unknown. V is commonly called the Robin potential.

Mathematically, the problem is to compute the integral

$$C = \int_{\Gamma} \mu(y)\, d\sigma(y)$$

and the function

$$\mu(y) = -\frac{1}{4\pi} \frac{\partial V}{\partial n}(y). \tag{6.1}$$

To find an equation satisfied by μ, we make use of the well-known jump properties of a single-layer potential's normal derivative. Hence, we have (for $x \in \mathbb{R}^3 \backslash G$ and $y \in \Gamma$, $|\cos \varphi_{yx}| = |\frac{y-x}{|x-y|} \cdot n(y)| \geq \alpha > 0$)

$$\begin{aligned} \frac{\partial V}{\partial n}(y) &= \lim_{x \to y} (\nabla_x V(x) \cdot n(y)) \\ &= -\int_{\Gamma} \frac{\cos \varphi_{yy'}}{|y - y'|^2} \mu(y') d\sigma(y') - 2\pi\mu(y). \end{aligned} \tag{6.2}$$

From (6.1) and (6.2), it follows that

$$\mu(y) = \int_{\Gamma} \frac{\cos \varphi_{yy'}}{2\pi|y-y'|^2} \mu(y') d\sigma(y'),$$

or in the operator notation

$$\mu = K\mu. \tag{6.3}$$

These equations imply that μ is the eigenfunction of the integral operator, K, corresponding to the maximal eigenvalue, which equals 1 [32].

Suppose that G is convex, then the kernel of the integral operator $k(y,y') = \frac{\cos \varphi_{yy'}}{2\pi|y-y'|^2}$ is non-negative and

$$\int_{\Gamma} \frac{\cos \varphi_{yy'}}{2\pi|y-y'|^2} d\sigma(y) = 1.$$

This normalization means that the kernel of the adjoint operator, K^*, $k(y_{n+1}, y_n) = r(y_n \to y_{n+1})$ can be considered as a transition probability density of a homogeneous Markov chain $\{y_n\}_{n=1}^{\infty}$ of points on Γ. This density function corresponds to the uniform distribution of successive points, y_{n+1}, in the solid angle with vertex y_n. Therefore, the Markov chain we constructed is the 3D version of the isotropic random walk-on-boundary process defined in Section 3.2.

Hence, we can think of $\Omega(y', \varsigma) = \int_{\varsigma} k(y,y') d\sigma(y)$ as a probability measure defined for every open set ς in Γ. This measure is based on choosing points uniformly in a solid angle subtended by ς and having y' as its vertex.

If Γ is strictly convex, then the angle measure, Ω, and surface measure, σ, are absolutely continuous, meaning that $\Omega(y', \varsigma)$ is strictly positive [26]. It is well known [47] that the weakly singular integral operator, K, is completely continuous. Thus, $\Omega(y', \varsigma)$ is regular and we can apply the ergodic theorem to statistics created using this Markov chain [26].

Suppose now that there are planar segments of the boundary. In this case, not every set of non-zero surface area $\sigma(\varsigma)$ has non-zero angle measure $\Omega(y', \varsigma)$ for all points $y' \in \partial G$. However, it can be easily shown that the second iteration, K^2, of the integral operator defines a strictly positive measure $\Omega^{(2)}(y', \varsigma) = \int_{\partial G} \Omega(y, \varsigma) k(y, y') d\sigma(y)$. Thus, we can apply the ergodic theorem in this case as well.

Therefore, by this theorem, there exists a positive stationary distribution, Π_∞, of the Markov chain as defined above. This means that

$$\Pi_\infty(\varsigma) = \int_{\Gamma} \Pi_\infty(d\sigma(y'))\, \Omega^{(2)}(y', \varsigma)$$

for every open set $\varsigma \subset \Gamma$. This also implies that the distribution is absolutely continuous and its density, π_∞, satisfies the equation

$$\pi_\infty(y) = K^2 \pi_\infty(y).$$

Also, the existence of $\pi_\infty(x)$ can be derived using the Döblin condition [55]. This condition is satisfied since Γ is an ergodic class.

Hence, since $\mu = K\mu = K^2\mu$,

$$\mu = C\pi_\infty \tag{6.4}$$

for some constant C. This constant must equal the capacitance of G, since π_∞ is a probability density. Here, we made use of the fact that 1 is a simple eigenvalue of the integral operator K, and there is only one eigenfunction corresponding to this eigenvalue [34].

By the ergodic theorem [38, 68], for an arbitrary initial distribution, Π_0, and bounded function v,

$$I[v] \equiv \int_\Gamma v(y)\pi_\infty(y)d\sigma(y) = \lim_{N\to\infty} \frac{1}{N}\sum_{n=1}^{N} v(y_n). \tag{6.5}$$

Note that we use both even and odd indexed points of the Markov chain in this sum since by (6.4), $\pi_\infty = K\pi_\infty$.

6.1.1 Monte Carlo estimator for computing capacitance

Consider the Robin potential, V, inside G. The boundary conditions state that the Robin potential is constant and equal to 1 in G. Thus, we have

$$\int_\Gamma \frac{1}{|x-y'|}\mu(y')d\sigma(y') = 1, \tag{6.6}$$

for any point $x \in G$. Therefore, we may fix $x \in G$ and set $v(y) = \frac{1}{|x-y|}$.

Together, relations (6.4), (6.5), (6.6) result in the following formula [104]:

$$C = \left(\lim_{N\to\infty} \frac{1}{N}\sum_{n=1}^{N} v(y_n) \right)^{-1}, \tag{6.7}$$

which will be used to calculate the capacitance.

To estimate the computational error, we use a Markov chain version of the central limit theorem [38, 68]. It states that $\frac{1}{N}\sum_{n=1}^{N} v(y_n)$ tends to a normally distributed random variable with mean $I[v]$ and variance $\sigma^2 N^{-1}$. Here,

$$\sigma^2 = \lim_{N\to\infty} \int_{\Gamma} \pi_\infty \left[\frac{1}{\sqrt{N}} \sum_{n=1}^{N} (v(y_n) - I[v]) \right]^2 .$$

This means that error of our computational algorithm is of the statistical nature. Hence, for a given accuracy ε, the cost of computations is of order $\sigma^2 \varepsilon^{-2}$.

To evaluate σ^2, we use the method of batch means [11] with the number of batches, $k+1$, equal to $\sqrt{N}+1$ and the batch size, m, equal to $\sqrt{N}$. Thus, we have

$$\sigma^2 = \lim_{m\to\infty,\ k\to\infty} \frac{m}{k} \sum_{i=0}^{k} \left(\frac{1}{m} S_i - \frac{1}{N} S \right)^2 ,$$

where $S_i = \sum_{j=mi+1}^{m(i+1)} v(y_j)$, $S = \sum_{i=0}^{k} S_i$.

Note that the algorithm based on (6.7) provides a method to obtain the value of the capacitance without explicitly calculating the density, μ.

6.1.2 Computing charge density

To calculate the charge distribution, we use relations (6.4) and (6.5) and construct Monte Carlo estimators for iterations either of the integral operator K or its adjoint, K^*. These estimators are based on a random walk-on-boundary process that is not necessarily isotropic. Let $p(y_n \to y_{n+1})$ be the transition probability density of this Markov chain, $\{y_n, n = 0, 1, \ldots\}$. Then, for some integrable functions $f \in L(\Gamma)$ and $h \in L^*(\Gamma)$, the direct and adjoint estimators, respectively, are defined as [104]

$$(h, K^n f) = \mathbb{E} Q_n h(y_n) = \mathbb{E} Q_n^* f(y_n).$$

Here

$$Q_0 = \frac{f(y_0)}{p_0(y_0)}, \quad Q_{n+1} = Q_n \frac{k(y_{n+1}, y_n)}{p(y_n \to y_{n+1})},$$

and

$$Q_0^* = \frac{h(y_0)}{p_0(y_0)}, \quad Q_{n+1}^* = Q_n^* \frac{k(y_n, y_{n+1})}{p(y_n \to y_{n+1})}.$$

Therefore, since we are integrating an absolutely convergent series, we have

$$(h, \mu) = C(h, \pi_\infty) = \lim_{N\to\infty} \frac{C}{N} \sum_{n=1}^{N} (h, K^n \pi_0). \tag{6.8}$$

It is clear that to compute the density, $\mu(y)$, at some point, $y \in \Gamma$, we have to set $h(y_0) = \delta(y - y_0)$. Note, however, that this last equality is valid only for bounded functions, h. To overcome this, we introduce a partition, $\varsigma_j, j = 1, \dots, m$, on the boundary surface: $\Gamma = \bigcup_{j=1}^{m} \varsigma_j$, and use a piecewise constant approximation for μ. Thus, we reduce the problem to the estimation of a finite number of cell values or, in other words, to a finite number of functionals (6.8) with different weight functions $h_j(y) = \chi[\varsigma_j](y)/\sigma(\varsigma_j)$. Here $\chi[\cdot]$ is the indicator function.

Hence, it is possible to use a direct estimator, and set $p_0 = f = \pi_0$. From this, it follows that for convex G, all weights, Q_n, are equal to 1 and so

$$\mu(y) \approx \lim_{N\to\infty} \frac{N_j C}{N\sigma(\varsigma_j)} \quad \text{for} \quad y \in \varsigma_j, \tag{6.9}$$

where N_j is the number of Markov chain points that hit the cell ς_j [104].

We thus arrive at an algorithm that makes it possible to calculate both the capacitance and the charge distribution simultaneously. Initially, we randomly choose a point y_0 on Γ with a probability density π_0. One of the possible choices for such a density is to set $\pi_0(y) = \frac{\cos\varphi_{yx}}{2\pi|y-x|^2}$ for some x inside the domain G. This means that y_0 is distributed isotropically within the solid angle with vertex x. Next, we simulate a long Markov chain of points using isotropic random walk on boundary and calculate C^{-1} using (6.7), and the numbers, N_j, with the methods described above. Finally, using (6.9), we obtain an approximation to the charge distribution.

6.2 Stationary diffusion equation with absorption

In this section, we deal with the following BVP (in $\mathbb{R}^3$, $\lambda \geq 0$) :

$$\begin{aligned} \Delta u(x) - \lambda u(x) = 0, \quad x \in G,\\ u|_\Gamma = g. \end{aligned} \tag{6.10}$$

Its fundamental solution, which turns to zero at infinity, is

$$Z(x-y|\lambda) = -\frac{1}{4\pi}\frac{1}{|x-y|}\exp\{-\sqrt{\lambda}|x-y|\}.$$

Hence, we can use the double-layer potential representation (for the **external** normal vector, n_ξ)

$$u(x) = \int_\Gamma \mu(\xi)\frac{\cos(\xi - x, n_\xi)}{2\pi|x-\xi|^2}\exp\{-\sqrt{\lambda}|x-\xi|\}(1+\sqrt{\lambda}|x-\xi|)d\sigma(\xi). \tag{6.11}$$

Here, the kernel equals $-2\frac{\partial Z(x-\xi|\lambda)}{\partial n_\xi}$ and μ is determined from

$$\mu(y) = -\int_\Gamma r(y\to\xi)h(y,\xi)\mu(\xi)d\sigma(\xi) + g(y). \tag{6.12}$$

In (6.12), the known functions are

$$h(x,\xi) = \exp\{-\sqrt{\lambda}|x-\xi|\}(1+\sqrt{\lambda}|x-\xi|)\, q(x,\xi)\, \mathrm{sign}(\cos(\xi - x, n_\xi)) \tag{6.13}$$

and

$$r(x \to \xi) = \frac{|\cos(\xi - x, n_\xi)|}{2\pi |x-\xi|^2 q(x,\xi)}, \tag{6.14}$$

where $q(x,\xi)$ is the number of intersections of the boundary by the line $\xi - x$.

Thus, we construct the Markov chain with the transition density (6.14) (the same function can be used to simulate the initial point of the Markov chain). Note that for sufficiently large λ, the kernel in (6.12) is sub-stochastic. Hence, an unbiased estimator can be constructed if, e.g.

$$\sup_{x,\xi\in\Gamma} |h(x,\xi)| < 1. \tag{6.15}$$

Condition (6.15) is satisfied if

$$\exp(-\theta)(1+\theta) < \frac{1}{q_{\max}},$$

where $\theta = \sqrt{\lambda} d$, d is the domain diameter and

$$q_{\max} = \sup_{x,\xi\in\Gamma} q(x,\xi).$$

For other variants of using the estimator for a non-convex boundary, see the remark in Section 3.7.1.

6.3 Multiply connected domains

In this section, we treat BVPs for multiply connected domains, in which the potential theory is more complicated than in the simply connected case. Following the book [34], we consider three cases:

(A) G is a bounded simply connected (1-connected) domain,

(J) G is a bounded domain, which is represented as

$$G = G_0 \setminus \bigcup_{l=1}^{k} G_l, \quad G_i \bigcap G_j = \emptyset, \quad i \neq j,$$

where all the domains $\{G_i\}_{i=0}^k$ are bounded, simply connected, and $G_i \subset G_0$, $i = 1, \dots, k$. Thus, the boundary of the domain G is $(k+1)$-connected:

$$\Gamma = \bigcup_{i=0}^{k} \Gamma_i.$$

(H) G is unbounded:

$$G = \mathbb{R}^3 \setminus \bigcup_{l=1}^{k} G_l, \quad G_i \bigcap G_j = \emptyset, \quad i \neq j.$$

In this section, we denote by n the normal vector **exterior** to G.

In Chapter 3, we constructed the walk-on-boundary algorithms for solving the Dirichlet, Neumann and third BVPs (interior and exterior), while in Section 6.1 we described the algorithm of solving the Robin problem for the case (A). We shall generalize these algorithms to the case (J) and (H).

In the case (J), as in (A), the condition

$$\int_{\Gamma} g(y) d\sigma(y) = 0 \tag{6.16}$$

is necessary to have the unique solution of the interior Neumann problem:

$$\begin{aligned} &\Delta\varphi(x) = 0, \ x \in G, \\ &\frac{\partial\varphi}{\partial n} \to g(y), \ \text{as} \ y \to \Gamma. \end{aligned} \tag{6.17}$$

In the case (J), the resolvent of the boundary integral equation for the density of the double-layer potential has, in general, two poles: $\lambda = 1$ and $\lambda = -1$. If (6.16) is satisfied, the pole $\lambda = -1$ disappears. Thus, the radius of convergence of the series representing the single-layer potential

$$\begin{aligned} &W(x) = -\frac{1}{2\pi} \int_{\Gamma} \frac{\mu(y) d\sigma(y)}{|x-y|}, \\ &\frac{\partial W}{\partial n} = -g, \end{aligned} \tag{6.18}$$

is equal to 1.

Let $V(x) = -W(x)$. Then,

$$W = V_1 + (-\lambda) V_2 + (-\lambda)^2 V_3 + \ldots + (-\lambda)^{k-1} V_k + \ldots, \tag{6.19}$$

where

$$V_k = -\frac{1}{2\pi} \int_{\Gamma} \frac{\mu_{k-1}(y) d\sigma(y)}{|x-y|}, \tag{6.20}$$

$$\mu_k(y) = -\frac{1}{2\pi} \int_{\Gamma} \mu_{k-1}(y') \frac{\cos\varphi_{yy'}}{|y-y'|^2} d\sigma(y'). \tag{6.21}$$

The multiplication method (we have already used it in Chapter 3) gives

$$W = -\frac{1}{2}\{V_1 + (V_2 + V_1) + (V_3 + V_2) + \ldots + (V_n + V_{n-1}) + \ldots\}. \tag{6.22}$$

This expansion is then calculated by the walk-on-boundary algorithm as described in Chapter 3.

In the case (E), the interior Neumann problem has a solution if

$$\int_{\Gamma_l} g(y)d\sigma(y) = 0, \quad l = 1, 2, \ldots, k. \tag{6.23}$$

If this condition is satisfied, the poles $\lambda = 1$, $\lambda = -1$ both disappear. Hence, the radius of convergence of series (6.19) is larger than 1. Taking in this series $\lambda = -1$, we get the solution to the interior BVP:

$$V = -\sum_{j=1}^{\infty} V_j.$$

Consider the exterior Neumann problem. The case (E) is treated exactly as the case (A); hence,

$$V = \frac{1}{2}\{V_1 - (V_2 - V_1) + (V_3 - V_2) + \ldots + (-1)^{n-1}(V_n - V_{n-1}) + \ldots\}. \tag{6.24}$$

In the case (J), the function W has, generally, two poles: $\lambda = 1$, $\lambda = -1$. However, condition (6.23) implies that the pole $\lambda = 1$ disappears; hence, in this case, (6.24) is also true.

Let us consider now the Robin problem.

The case (J) is treated as the case (A) described in Section 6.1. Indeed, $\mu(y) = 0$ on Γ_l, $l = 1, \ldots, k$. Thus, the function μ is defined on Γ_0.

The most complicated situation with the Robin problem arises in the case (E). Indeed, for $\lambda = -1$, we have k different eigenfunctions satisfying the conditions:

$$\mu^{(l)} = -K^* \mu^{(l)}, \quad l = 1, \ldots, k,$$
$$\int_{\Gamma_i} \mu^{(l)}\, d\sigma = \delta_{il}, \tag{6.25}$$

where δ_{il} is the Kronecker symbol, $i, l = 1, \ldots, k$.

Let

$$\int_{\Gamma_l} \frac{\partial V}{\partial n}\, d\sigma = \alpha^{(l)}$$

be the 'charge' on the surface Γ_l, $l = 1, \dots, k$. The solution to the Robin problem is fully defined by the expression

$$\sum_{i=1}^{k} \alpha^{(i)} \mu^{(i)},$$

where $\mu^{(i)}$, $i = 1, \dots, k$ are found from (6.25).

Thus, we have to solve the problem of finding k eigenfunctions of the operator K^* corresponding to the eigenvalue $\lambda = -1$. Here, we need the information on the geometrical multiplicity of the pole $\lambda = 1$. The following result is known [42]. Consider the general case when the domain consists of p non-overlapping domains $\Omega_1, \dots, \Omega_p$ with cuts $\Phi_1, \dots, \Phi_q$ (q is the total number of cuts).

Proposition 6.1. *The geometric multiplicity of the eigenvalue $\lambda = -1$ is equal to p. Functions $\theta_j \in C(\Gamma)$ defined by $\theta_j = 1$ on $\partial\Omega_j$ and $\theta_j = 0$ on $\partial\Omega_k$ for all $k \neq j$ represent the basis of the eigenspace of the operator K (i.e., of the space $N(I - K)$).*

For $q > 0$, $\lambda = 1$ is an eigenvalue of K with the geometrical multiplicity q. The functions $\theta_j \in C(\Gamma)$ defined by $\theta_j = 1$ on $\partial\Phi_j$ and $\theta_j = 0$ on $\partial\Phi_k$ for all $k \neq j$ represent the basis of the eigenspace $N(I - K)$.

Thus, for the case (E), we take $p = k$, $q = 0$. To satisfy the condition (6.25), we choose $\theta_j = 1/S_j$ on $\partial\Omega_j$ and $\theta_j = 0$ on $\partial\Omega_l$ ($l \neq j$), $j = 1, \dots, k$, where $S_j = \sigma(\partial\Omega_j)$ is the area of the surface $\partial\Omega_j$.

Write down the expansion of the resolvent $R(\lambda, K^*)$ into the Laurent series in the neighbourhood of the pole $\lambda = \lambda_j = -1$:

$$R(\lambda, K^*) = \sum_{k=0}^{\infty} (\lambda - \lambda_j)^k \, T_{kj} + \sum_{k=1}^{\infty} (\lambda - \lambda_j)^{-k} \, B_{kj} \tag{6.26}$$

and suppose that

$$B_{sj}\theta_j \neq 0, \quad B_{s+1,j}\theta_j \neq 0, \tag{6.27}$$

where θ_j, $j = 1, \dots, k$ are particular initial densities. Let

$$\mu_{jn} = \frac{1}{n} \sum_{i=1}^{n} i^{-s+1} \lambda_j^{-i} (K^*)^i \theta_j, \quad \lambda_j = -1, j = 1, \dots, k. \tag{6.28}$$

Proposition 6.2. *Sequence (6.28) converges in norm to the eigenvector $\mu^{(j)}$, which corresponds to the eigenvalue λ_j, (i.e., $K^* \mu^{(j)} = \lambda_j \mu^{(j)}$), if the finite number of poles of the resolvent $R(\lambda, K^*)$ lies on the boundary of the spectrum of the operator K^*.*

This result is proved [63] in the general case, when the poles $\lambda_1, \ldots, \lambda_k$ have the multiplicities $q_1, \ldots, q_k$, respectively. In this case, it is supposed in (6.27) that $s \geq q_j$, $j = 1, \ldots, k$.

Thus, to find numerically the eigenfunctions $\{\mu^{(i)}\}_{i=1}^k$, we can use the sequence

$$\mu^{(j)} \approx \mu_{jn} = \frac{1}{n} \sum_{i=1}^{n} i^{-k} (-1)^{-i} (K^*)^i \theta_j, \quad j = 1, \ldots, k, \tag{6.29}$$

where the iterations $(K^*)^i \theta_j$ are calculated by the walk-on-boundary algorithm, as described for the non-convex case. The rate of convergence of the series is determined by the quantity $\ln(n)/n$.

To complete the considerations of the Robin problem, it remains to study the case when then pole is $\lambda = 1$ (this appears in the case (J) for the operator K^*).

In the case (J), there are k eigenfunctions (Statement 6.1):

$$\nu^{(j)} = K^* \nu^{(j)}, \quad j = 1, \ldots, k.$$

Since there are only two simple poles $\lambda = 1$ and $\lambda = -1$, the radius of convergence of the series

$$(1 - \lambda_2)\nu = \nu_1 + \lambda_2 + \lambda^2(\nu_2 - \nu_0) + \ldots + \lambda^n(\nu_n - \nu_{n-2}) + \ldots$$

is larger than 1. Here, the coefficients $\{\nu_n\}_{n=0}^{\infty}$ are defined by the expansion:

$$\nu = \nu_0 + \lambda \nu_1 + \lambda^2 \nu_2 + \ldots,$$

hence,

$$\nu_0 = g, \; \nu_1 = -K^* g, \; \ldots, \; \nu_n = -K^* \nu_{n-1}.$$

This implies that the sequences $\nu_0, \nu_2, \ldots$, and $\nu_1, \nu_3, \ldots$, converge to limits, say H_e and H_0, respectively. Then [34] $H_e - H_0$ is an eigenfunction corresponding to the eigenvalue $\lambda = 1$, and

$$\int_{\Gamma_l} (H_e - H_0)\, d\sigma = 2 \int_{\Gamma_l} g\, d\sigma.$$

This equality shows that it is possible to find k functions of the type H_e and H_0 depending on the functions g_l:

$$\begin{aligned} g_l(x) &= 0, \quad x \notin \Gamma_l, \\ g_l(x) &= \frac{1}{2S^{(l)}}, \quad x \in \Gamma_l, l = 1, \ldots, k. \end{aligned} \tag{6.30}$$

Let $v^{(l)} = H_e^{(l)} - H_0^{(l)}$. Then,

$$\int_{\Gamma_m} v^{(l)}\, d\sigma = 0, \quad l \neq m;$$

$$\int_{\Gamma_l} v^{(l)}\, d\sigma = 1; \quad \int_{\Gamma_m} g^{(l)}\, d\sigma = \frac{1}{2}. \tag{6.31}$$

An arbitrary choice of g generates an eigenfunction of the form

$$v = \alpha^{(1)} v^{(1)} + \ldots + \alpha^{(k)} v^{(k)},$$

where

$$\alpha^{(l} = 2 \int_{\Gamma_l} g\, d\sigma, \quad l = 1, \ldots, k.$$

Consider now the interior Dirichlet problem. In the general case of a multiply connected domain, the walk-on-boundary algorithm presented in Chapter 3 does not work since the spectral properties of the integral operator are different. However in the case (E) it works, since the function

$$W = W_1 + \lambda W_2 + \lambda^2 W_3 + \ldots \tag{6.32}$$

may have only the pole $\lambda = -1$. In (6.32),

$$W_n = \frac{1}{2\pi} \int_{\Gamma} \frac{\cos \varphi_{yy'}}{|y - y'|^2} \mu_{n-1}(y')\, d\sigma(y'), \quad n = 1, \ldots,$$

$$\mu_0 = g, \ \ldots, \ \mu_n = \frac{1}{2\pi} \int_{\Gamma} \frac{\cos \varphi_{y'y}}{|y - y'|^2} \mu_{n-1}(y')\, d\sigma(y'), \ n = 1, \ldots. \tag{6.33}$$

Thus, (6.32) gives the representation for the double-layer potential

$$W(x) = \frac{1}{2\pi} \int_{\Gamma} \mu(y) \frac{\cos \varphi_{yx}}{|x - y|^2}\, d\sigma(y)$$

with the boundary condition $W|_{\Gamma} = g$.

To calculate (6.32) at $\lambda = 1$, we use the multiplication method described in Chapter 3 (multiplying by $(\lambda + 1)$). This gives

$$W = \frac{1}{2}[W_1 - (W_2 - W_1) + (W_3 - W_2) - \ldots], \quad W|_{\Gamma} = g. \tag{6.34}$$

Let us consider now the interior Dirichlet problem for the case (J). The solution of the BVP

$$\Delta V = 0, \quad V|_{\Gamma} = g$$

is constructed in the form $V = W + V_0$, where W is the double-layer potential and V_0 is the single-layer potential. It is clear that the solution cannot be found in the form of only double-layer potential since $\lambda = 1$ is also a pole of W. Although the integral equation $\mu = \lambda K\mu + g$ may, generally, have $\lambda = 1$ as a pole, it is possible to impose some conditions that lead to disappearance of this pole. These conditions are

$$\int_\Gamma g v^{(l)}\, d\sigma = 0, \quad l = 1, \dots, k, \tag{6.35}$$

where $v^{(l)}$, $l = 1, \dots, k$, are the eigenfunctions of the equation $v = \lambda K^* v$ corresponding to $\lambda = 1$ (we discussed the construction of these eigenfunctions above, see (6.31)). It is clear that these conditions are artificial for the Dirichlet problem. If in some special case these conditions were satisfied, we could use representation (6.34), since only the pole $\lambda = -1$ could exist.

Thus, we consider the general case that does not require conditions (6.35). As mentioned, we seek the solution as a sum of double- and single-layer potentials. Let us define a function $\alpha(x)$, $x \in \Gamma$: $\alpha = 0$ for $x \in \Gamma_0$, and $\alpha = \alpha^{(l)}$ on Γ_l, $l = 1, \dots, k$; here, $\alpha^{(l)}$ are some constants, which we choose so that the function $g - \alpha$ satisfies (6.35). From this, we get

$$\alpha^{(l)} = \int_\Gamma g v^{(l)}\, d\sigma. \tag{6.36}$$

We can find coefficients $\gamma_1, \dots, \gamma_k$ so that on Γ_j, $j = 1, \dots, k$, the following equalities are true:

$$\sum_{l=1}^{k} \gamma_l \int_\Gamma \frac{v^{(l)}(y)\, d\sigma(y)}{|x - y|} = \alpha^{(j)}. \tag{6.37}$$

Let

$$V_0 = \int_\Gamma \sum_{l=1}^{k} \gamma_l v^{(l)} \frac{d\sigma(y)}{|x - y|}. \tag{6.38}$$

Let $W(x)$ be a double-layer potential with the density μ satisfying the equation

$$\mu(y) = \lambda K\mu + g - \alpha(y), \quad \lambda = 1. \tag{6.39}$$

In this case, $W(x)$ can be found from (6.3), but $W|_\Gamma = g - \alpha$. Thus, $V = W + V_0$, $V|_\Gamma = g - \alpha + \alpha = g$.

Thus, we first calculate $\alpha^{(l)}$ in (6.36) on the basis of (6.29). Then, from (6.37), we find the coefficients $\gamma_1, \dots, \gamma_k$. The linear functionals in (6.38) are calculated from the approximation (6.28), which in our case takes the form

$$v^{(j)} \approx v_{jn} = \frac{1}{n}\sum_{i=1}^{n}(i)^k (K^*)^i g_j.$$

We turn now to the exterior Dirichlet problem.

The case (E) (including $k = 1$, i.e., the case (A)).

The situation is the same as in the case (J) of the interior Dirichlet problem. We first write down the conditions, which ensures that $\lambda = -1$ is not a pole of the density μ determined from

$$\mu = \lambda K\mu - g. \tag{6.40}$$

Since we have already treated the Robin problem for the case (E) and have found k eigenfunctions from (6.25) at $\lambda = -1$, we can write the condition which ensures that (6.40) has a solution for $\lambda = -1$:

$$\int_\Gamma g\mu^{(l)}\, d\sigma = 0, \quad l = 1, \dots, k. \tag{6.41}$$

In the case (A), $k = 1$ and there is one eigenfunction $\mu^{(1)}$. Consequently, if (6.41) is satisfied, the functions $\mu^{(l)}$ have no pole at $\lambda = -1$. But $\lambda = 1$ is also not a pole of $\mu^{(l)}$; hence, the Neumann series $W = W_1 + W_2 + \dots$ is convergent.

However, conditions (6.41) are artificial, and we use the same approach, constructing the solution in the form of a sum of double- and single-layer potentials.

Using the solvability conditions, we choose the function α (as $\alpha^{(l)} = \text{const}$ on Γ_l) so that

$$\alpha^{(l)} = \int_\Gamma g\mu^{(l)} d\sigma \quad l = 1, \dots, k \tag{6.42}$$

and the coefficients $\gamma_1, \dots, \gamma_k$ are found from

$$\sum_{l=1}^{k}\gamma_l \int_\Gamma \frac{\mu^{(l)}(y)\, d\sigma(y)}{|x-y|} = \alpha^{(j)} \tag{6.43}$$

on Γ_j, $j = 1, \dots, k$. Let

$$V_0 = \int_\Gamma \sum_{l=1}^{k}\gamma_l \mu^{(l)} \frac{d\sigma(y)}{|x-y|}. \tag{6.44}$$

Then, $V = V_0 - W$ is the solution of our problem, since $V|_\Gamma = \alpha - \alpha + g = g$. Linear functionals of the eigenfunctions can be calculated by the scheme (6.29).

The exterior Dirichlet problem for the case (J).

This problem is equivalent to the problem of finding the solution to the exterior Dirichlet problem for the domain with the surface Γ_0 and k interior Dirichlet problems in domains with the surfaces Γ_l, $l = 1, \ldots, k$. However, we can treat the problem using the general considerations. In this case, we have two poles: $\lambda = -1$ and $\lambda = 1$ whose geometrical multiplicity is 1 and k, respectively. Hence, the equation adjoint to (6.40) has one eigenfunction $\mu^{(1)}$ (we studied this function when constructing the solution to the Robin problem for the case (A)). Consequently, (6.40) has a solution for $\lambda = -1$, if

$$\int_\Gamma g\mu^{(1)}\, d\sigma = 0. \tag{6.45}$$

Thus, if this condition is satisfied, W has only one pole $\lambda = 1$; therefore, the multiplication method gives

$$W = \frac{1}{2}[W_1 + (W_2 + W_1) + (W_3 + W_2) + \ldots]. \tag{6.46}$$

Conditions (6.45) are artificial for the problem under study. Suppose that

$$\int_\Gamma \mu^{(1)}\, d\sigma = 1.$$

The potential

$$C = \int_\Gamma \frac{\mu^{(1)}\, d\sigma}{|x-y|}$$

is constant inside the domain G. Let

$$\alpha = \int_\Gamma g\mu^{(1)}\, d\sigma.$$

We construct w, the solution of the Dirichlet problem with the boundary conditions: $g' = g - \alpha$. This function satisfies the condition (6.45); therefore,

$$w = \frac{1}{2}[w_1 + (w_2 + w_1) + (w_3 + w_2) + \ldots], \quad w|_\Gamma = g,$$

where

$$w_1 = \frac{1}{2\pi}\int_\Gamma (g-\alpha)\frac{\cos\varphi_{yy'}}{|y-y'|^2}\, d\sigma(y'),$$

and

$$w_k = \frac{1}{2\pi}\int_\Gamma w_{k-1}\frac{\cos\varphi_{yy'}}{|y-y'|^2}\, d\sigma(y').$$

Therefore, $w_k = W_k$ in G_l and $W|_\Gamma = \alpha - g$.

Let

$$V_0 = \frac{\alpha}{C} \int_\Gamma \frac{\mu^{(1)}\, d\sigma}{|x-y|},$$

and

$$V = V_0 - W. \tag{6.47}$$

This function solves the problem, since

$$V|_\Gamma = \frac{\alpha C}{C} - \alpha + g = g.$$

The linear functionals of $\mu^{(1)}$ such as α, C and

$$\int_\Gamma \frac{\mu^{(1)}\, d\sigma}{|x-y|}$$

can be calculated by the algorithm described in Section 6.1.

In conclusion, we summarize all the situations we considered in this section.

(1) The exterior Neumann problem

$$\Delta\varphi = 0, \quad \left.\frac{\partial\varphi}{\partial n}\right|_\Gamma = g^*.$$

The boundary integral equation for the density of the single-layer potential is

$$\mu^* = \lambda K^* \mu^* + g^*; \quad (\lambda = 1). \tag{6.48}$$

(2) The interior Neumann problem

$$\Delta\varphi = 0, \quad \left.\frac{\partial\varphi}{\partial n}\right|_\Gamma = g^*.$$

The boundary integral equation for the density of the single-layer potential is

$$\mu^* = \lambda K^* \mu^* - g^*; \quad (\lambda = -1). \tag{6.49}$$

(3) The interior Dirichlet problem

$$\Delta\varphi = 0, \quad \varphi|_\Gamma = g.$$

The boundary integral equation for the density of the double-layer potential is

$$\mu = \lambda K\mu + g; \quad (\lambda = 1). \tag{6.50}$$

(4) The exterior Dirichlet problem

$$\Delta\varphi = 0, \quad \varphi|_\Gamma = g.$$

The boundary integral equation for the density of the double-layer potential is

$$\mu = \lambda K\mu - g; \quad (\lambda = -1). \tag{6.51}$$

(5) The eigenvalue problem for the operator K^* at $\lambda = -1$:

$$\mu^{(l)} = -K^*\mu^{(l)}. \tag{6.52}$$

The geometrical multiplicity of the pole $\lambda = -1$ is equal to 1 in cases (A) and (J); the corresponding eigenfunction $\mu^{(1)}$ solves the Robin problem. In the case (E), the geometrical multiplicity of the pole $\lambda = -1$ is equal to k (k is the number of components of the domain). The corresponding eigenfunctions $\{\mu^{(l)}\}_{l=1}^k$ can be found from (6.29).
When solving the exterior Dirichlet problem in cases (A) and (J), only the function $\mu^{(1)}$ used, while in the case (E) all functions $\{\mu^{(l)}\}_{l=1}^k$ are involved, since the equation (6.49) is adjoint to (6.51); This leads to the representation of the solution in the form (6.47) (cases (A) and (J)) or (6.42),(6.44) (case (E)).

(6) The eigenvalue problem for the operator K^* at $\lambda = 1$:

$$\mu^{(l)} = K^*\mu^{(l)}. \tag{6.53}$$

There is a pole $\lambda = 1$ only in the case (J). The geometrical multiplicity of the pole is equal to k, the number of cuts in the domain. Since the equation (6.48) is adjoint to (6.50), then, when solving the interior Dirichlet problem in the case (J), the functions $\{\nu^{(l)}\}_{l=1}^k$ are used.

6.4 Stabilization method

To extend the walk-on-boundary algorithms to multiconnected domains, there exists a different approach. The algorithms constructed in Chapter 4 for the heat equation are valid in this general case. We can use these algorithms and then apply the stabilization method.

Consider the following pair of equations in a bounded domain:

$$\frac{\partial u}{\partial t}(x,t) = \Delta u(x,t) + f(x), \quad x \in G,\ t > 0,$$

$$u(x,0) = 0, \quad x \in G,$$

$$u(y,t) = \psi(y), \quad y \in \partial G,\ t > 0$$

and

$$\Delta v(x) + f(x) = 0, \quad x \in G,$$

$$v(y) = \psi(y), \quad y \in \partial G.$$

The function $\bar{u}(x,t) = u(x,t) - v(x)$ solves the problem:

$$\frac{\partial \bar{u}}{\partial t}(x,t) = \Delta \bar{u}(x,t), \quad x \in G,\ t > 0,$$

$$\bar{u}(x,0) = -v(x), \quad x \in G,$$

$$\bar{u}(y,t) = 0, \quad y \in \partial G,\ t > 0.$$

Denote by $\{\lambda_k\}_{k=1}^{\infty}$ the set of eigenvalues and by $\{\varphi_k\}_{k=1}^{\infty}$ the eigenfunctions of the Laplace operator:

$$\Delta \varphi_k + \lambda_k \varphi_k = 0, \quad x \in G,$$

$$\varphi_k(y) = 0, \quad y \in \partial G, k = 1, 2, \ldots$$

and λ_1 is the minimal eigenvalue. From the representation

$$\bar{u}(x,t) = \sum_{k=1}^{\infty} c_k \exp(-\lambda_k t)\ \varphi_k(x),$$
$$\Delta \bar{u}(x,t) = \sum_{k=1}^{\infty} d_k \exp(-\lambda_k t)\ \varphi_k(x),$$

where c_k, d_k are the Fourier coefficients in the expansions of $v(x)$ and $f(x)$, respectively, we get the inequalities

$$||\bar{u}(\cdot,t)||_{L_2(G)} \le ||v||_{L_2(G)} \exp(-\lambda_1 t),$$
$$||\Delta \bar{u}(\cdot,t)||_{L_2(G)} \le ||f||_{L_2(G)} \exp(-\lambda_1 t).$$

Using the condition $\bar{u}(y) = 0$ for $y \in \partial G$ and the embedding theorem, we get the following estimation in the uniform norm:

$$||\bar{u}(\cdot,t)||_{C(G)} \le c \exp(-\lambda_1 t).$$

Consequently, if we solve the stationary problem by the stabilization method and the desired accuracy is defined by the inequality:

$$||\bar{u}(\cdot,t)||_{C(G)} \le \varepsilon,$$

then it is necessary to achieve the times $t \sim |\ln(\varepsilon)|$. Since $\mathbb{E}N_t \sim t$, and $\mathbb{D}\xi_{xt}^{(i)} \sim t^2$, we conclude that the cost of the stabilization method is

$$T_\varepsilon \sim |\ln(\varepsilon)|^3/\varepsilon^2.$$

6.5 Nonlinear Poisson equation

In contrast to the WOS method, the walk-on-boundary algorithms give the solution at an arbitrary set of points, say, in the points of a grid. This can be used to construct an effective implementation of the known iterative method [12].

Consider the following BVP in a bounded domain $G \subset \mathbb{R}^n$:

$$\begin{aligned} \Delta u &= f(x_1, x_2, \dots, x_n, u) = f(x, u), \\ u|_\Gamma &= \varphi, \end{aligned} \tag{6.54}$$

where it is assumed that

(1) f has continuous derivatives for all $x \in G \cup \Gamma$ and for all u;

(2) $|f(x, u)| \le N$.

Under these conditions, (6.54) has a solution that can be found by the following iterative process. In what follows, we suppose that $\varphi = 0$; this can be achieved by adding a harmonic function.

Let

$$k = \sup\left\{\frac{\partial f(x, u)}{\partial u}\right\}$$

for all $x \in G$.

Denote by v the solution to the BVP:

$$\begin{aligned} \Delta v &= -N, \\ v|_\Gamma &= 0. \end{aligned} \tag{6.55}$$

Consider the iterative process:

$$\begin{aligned} u_0 &= v, \\ \Delta u_m - k u_m &= f(x, u_{m-1}) - k u_{m-1} \equiv g(x, u_{m-1}), \\ u_m|_\Gamma &= 0. \end{aligned} \tag{6.56}$$

It is known [12] that $u_m \to u$ monotonically.

If G is small, the solution to (6.54) is unique. Generally, (6.56) gives the maximal solution to (6.54). Note that if we start with $u_0 = -v$, we arrive at the minimal solution.

Thus, it is necessary to successively solve the series of BVPs of the type:

$$\begin{aligned} \Delta u_m - k u_m &= g(x, u_{m-1}), \\ u_m|_\Gamma &= 0. \end{aligned} \tag{6.57}$$

Let $\mathcal{E}_k(x,y)$ be the fundamental solution to the equation

$$\Delta u - ku = \delta(x-y),$$

i.e.,

$$\mathcal{E}_k(x,y) = \frac{1}{2\pi}\frac{\exp[-\sqrt{k}|x-y|]}{|x-y|}.$$

We seek the solution to (6.57) in the form

$$u(x) = \int_G g(x,u) u_\delta(x,y)dy,$$

where

$$u_\delta(x,y) = \mathcal{E}_k(x,y) + w(x,y).$$

Hence, $w(x,y)$ solves the problem

$$\begin{aligned} (\Delta_x - k)w(x,y) &= 0, \\ w(x,y)|_{x\in\Gamma} &= -\mathcal{E}_k(x,y)|_{x\in\Gamma}. \end{aligned}$$

First, we write down the random estimator for the problem

$$\begin{aligned} \Delta u_m - k u_m &= 0, \\ u|_\Gamma &= \varphi. \end{aligned} \tag{6.58}$$

Let $\xi(x)$ be a random estimator for the Dirichlet problem

$$\begin{aligned} \Delta v &= 0, \\ v|_\Gamma &= \varphi. \end{aligned} \tag{6.59}$$

Then, the random estimator for (6.58) is

$$\eta(x) = Q\xi(x),$$

where

$$Q = (1 - \sqrt{k}|y_{i+1} - y_i|)\exp[\sqrt{k}|y_{i+1} - y_i|],$$

and y_{i+1}, y_i are two successive boundary points of the walk-on-boundary trajectory. Next, to evaluate the solution using its integral representation, we apply the double randomization principle.

Thus, the algorithm for solving the nonlinear problem (6.54) can be implemented according to the following scheme:

(1) First, store a set of points of the walk-on-boundary process for M trajectories. M should be sufficiently large to attain the desired accuracy.
(2) Calculate $u_0 = v$, the solution to (6.55), for all points of a grid chosen in G using the global walk-on-boundary algorithm described in Chapter 3. This is possible since $v|_\Gamma = 0$.
(3) Using the estimator η, perform the same calculation for (6.57). Note that in all estimators, one and the same set of points sampled in p.1 is used.

If $f(x, u)$ is a bounded function, we get a Monte Carlo estimator with a finite variance.

7 Splitting and survival probabilities in random walk methods and applications

In this chapter we present random walk–based methods for solving some practically interesting problems that are based on the probabilistic representations and different Wiener process properties. In particular, the first passage time and survival probabilities of the Wiener process are used in the problem of diffusion imaging and cathodoluminescence studies. We describe a series of extremely fast stochastic algorithms based on exact representations for the first passage time and exit point probability densities, splitting and survival probabilities. We discuss possible applications of the developed algorithms to the following problems: (1) simulation of epitaxial nanowire growth and (2) diffusion imaging of microstructures, in particular, cathodoluminescence imaging for threading dislocations. In the next chapter, we will use these ideas to simulate the annihilation of electrons and holes in vicinity of nonradiative centres and the quantum efficiency evaluation. In the last example, the random WOS method is used to solve nonlinear diffusion equations and more general systems of nonlinear Smoluchowski equations combined with the kinetic Monte Carlo (KMC) method.

7.1 Introduction

The probabilistic interpretations of PDEs and related stochastic algorithms are well developed mainly for parabolic and elliptic partial differential equations (e.g. see [19, 29, 98]). Due to deep intrinsic relation between diffusion random processes and the parabolic equations, the solutions to the relevant BVPs are represented in the form of expectations of functionals over trajectories of the random diffusion processes. A straightforward application of this approach is based on a numerical solution of stochastic differential equations governing the random diffusion processes (e.g. see [44, 52]).

Of particular interest are special cases where the solution to a PDE can be represented not through the whole continuous trajectory of the diffusion process but through much simpler random events only, like a survival probability in a domain, the first passage time, known as the exit time from a domain, splitting probabilities describing the probabilities to hit different parts of the boundary, the exit positions and many others. A well known and widely used is the random WOS method for solving linear PDEs with constant coefficients (e.g. see [24, 98]). This technique is extremely efficient when not all the whole solution field is desired but rather some integrals of the solution, or solutions at some prescribed points.

In this and the next chapter, we suggest different random walk–based algorithms for solving three classes of physical problems, in particular, (1) a diffusion tomography related to cathodoluminescence imaging technique, (2) simulation of epitaxial growth

$$
\begin{aligned}
&= \sum_{n=1}^{\infty}(-1)^{n+1}\frac{\pi n}{rR}\sin(\pi rn/R)\int_0^{\infty} dt\exp\left\{-t\left(\frac{\lambda^2}{2}+\frac{\pi^2 n^2}{2R^2}\right)\right\}\\
&= \sum_{n=1}^{\infty}(-1)^{n+1}\frac{\pi n}{rR}\sin(\pi rn/R)\left[\frac{2}{\lambda^2+\pi^2 n^2/R^2}\right]\\
&= \frac{\sinh(\lambda r)/\lambda r}{\sinh(\lambda R)/\lambda R}. \qquad (7.9)
\end{aligned}
$$

The last equality follows from the series representation (see [85], **8**, p. 731):

$$
\sum_{n=1}^{\infty}(-1)^{n+1}\frac{n}{n^2+a^2}\sin(nx) = \frac{\pi}{2}\sinh(ax)/\sinh(\pi a).
$$

One-dimensional case.

In what follows, we will need also the survival probability for the one-dimensional (1D) case. Let us consider an interval $[-R, R]$, and we are interested in finding the survival probability of a Wiener process starting from an arbitrary point $x \in [-R, R]$.

From the general expression for the first passage time, in 1D case, we have

$$
p(t,x,R) = \frac{\pi}{2R^2}\sum_{k=0}^{\infty}(-1)^k(2k+1)\cos\left[(2k+1)/2R\right]\exp\{-(2k+1)^2\pi^2 t/8R^2\}.
$$

Therefore,

$$
\begin{aligned}
\langle \exp\{-\lambda^2\tau/2\}\rangle &= \int_0^{\infty} dt\exp\{-\lambda^2 t/2\}p(t,r,R)\\
&= \int_0^{\infty} dt\exp\{-\frac{\lambda^2 t}{2}\}\frac{\pi}{2R^2}\sum_{k=0}^{\infty}(-1)^k(2k+1)\cos\left[(2k+1)/2R\right]\\
&\quad\times\exp\left\{-\frac{(2k+1)^2\pi^2 t}{8R^2}\right\}\\
&= \frac{4}{\pi}\sum_{k=0}^{\infty}(-1)^k(2k+1)\frac{\cos\left[(2k+1)/2R\right]}{(2k+1)^2+4R^2\lambda^2/\pi^2}\\
&= \frac{\cosh(\lambda x)}{\cosh(\lambda R)}
\end{aligned}
$$

since (see [85], **27**, p. 734)

$$
\sum_{k=0}^{\infty}(-1)^k\frac{2k+1}{(2k+1)^2+z^2}\cos\left[(2k+1)x\right] = \frac{\pi}{4}\frac{\cosh(xz)}{\cosh(\pi z/2)}.
$$

Two-dimensional case.

From the general expression for the first passage time, in 2D case, we have

$$p(t,r,R)=\sum_{n=1}^{\infty}\frac{\mu_n}{R^2}\frac{J_0(\mu_n r/R)}{J_1(\mu_n)}\exp\{-\mu_n^2 t/2R^2\},$$

where μ_n are the positive solutions of the equation $J_0(x)=0$.

Therefore,

$$\begin{aligned}
\langle\exp\{-\lambda^2\tau/2\}\rangle &= \int_0^{\infty} dt\,\exp\{-\lambda^2 t/2\}p(t,r,R)\\
&= \int_0^{\infty} dt\,\exp\{-\frac{\lambda^2 t}{2}\}\sum_{n=1}^{\infty}\frac{\mu_n}{R^2}\frac{J_0(\mu_n r/R)}{J_1(\mu_n)}\exp\{-\mu_n^2 t/2R^2\}\\
&= \sum_{n=1}^{\infty}\frac{\mu_n}{R^2}\frac{J_0(\mu_n r/R)}{J_1(\mu_n)}\int_0^{\infty}\exp\{-[t(\mu_n^2/2R^2+\lambda^2/2)]\}\,dt\\
&= \sum_{n=1}^{\infty}\frac{J_0(\mu_n r/R)}{J_1(\mu_n)}\left[\frac{2\mu_n}{R^2\lambda^2+\mu_n^2}\right]\\
&= \frac{I_0(\lambda r)}{I_0(\lambda R)},
\end{aligned} \tag{7.10}$$

where $I_0(x)$ is the zero-order modified Bessel function: $I_0(x)=J_0(ix)$. In the last equality in (7.10), we used the relation (see [86], **5.7.33, 4**, p. 690)

$$\sum_{n=1}^{\infty}\frac{\mu_n}{(\mu_n^2-a^2)}\frac{J_0(\mu_n x)}{J_1(\mu_n)}=\frac{J_0(ax)}{2J_1(a)},$$

which we apply for $a=i\lambda R$, $x=r/R$.

We will see later that when we deal with time-independent solutions, the survival and splitting probabilities can be evaluated much simpler using the reciprocity theorem.

7.3 The reciprocity theorem for particle collection in the general case of Robin boundary conditions

In this section, we prove a reciprocity theorem, which we use in the constructions of different stochastic algorithms.

Let Γ be a surface enclosing a volume where a diffusional particle collection occurs. The surface is composed of parts Γ_i $(i=1,\dots,N)$.

We consider the basic case of a unit source of particles at $\mathbf{r}'$. The distribution of the diffusing particles is governed by the diffusion equation

$$D\Delta G(\mathbf{r},\mathbf{r}') - \tau^{-1} G(\mathbf{r},\mathbf{r}') = -\delta(\mathbf{r}-\mathbf{r}'), \tag{7.11}$$

where D is the diffusion coefficient and τ is the particle's lifetime.

We first present the reciprocity theorem given in [98], which concerns the case when on the boundary Γ_1 where the particles flux

$$I(\mathbf{r}) = -D \int_{\Gamma_1} \frac{\partial G(\xi, r)}{\partial \nu_\xi}\, d\sigma(\xi) \tag{7.12}$$

is calculated, the vanishing boundary conditions are prescribed:

$$G(\mathbf{r},\mathbf{r}') = 0 \quad \text{for} \quad \mathbf{r} \in \Gamma_1, \tag{7.13}$$

while on the surfaces $\Gamma_2, \dots, \Gamma_N$, the Robin boundary conditions are posed:

$$-\frac{\partial G(\mathbf{r},\mathbf{r}')}{\partial \nu_i} = S_i G(\mathbf{r},\mathbf{r}') \quad \text{for} \quad \mathbf{r} \in \Gamma_i (i = 1, 2, \dots, N). \tag{7.14}$$

Here, S_i are reduced (i.e., divided by D), surface recombination velocities ($0 \le S_i \le \infty$).

The reciprocity theorem says that $I(\mathbf{r}) \equiv F(\mathbf{r})$ where $F(\mathbf{r})$ is the unique solution of the following homogeneous BVP:

$$D\Delta F(\mathbf{r}) - \tau^{-1} F(\mathbf{r}) = 0, \tag{7.15}$$

with the modified boundary condition on Γ_1:

$$F(\mathbf{r}) = 1 \quad \text{for} \quad \mathbf{r} \in \Gamma_1 \tag{7.16}$$

but with the same boundary conditions on other surfaces:

$$-\frac{\partial F(\mathbf{r})}{\partial \nu_i} = S_i F(\mathbf{r}) \quad \text{for} \quad \mathbf{r} \in \Gamma_i (i = 2, \dots, N). \tag{7.17}$$

The proof immediately follows from the Green formula

$$F(\mathbf{r}') = -D \int_{\Gamma} \left[F\frac{\partial G}{\partial \nu} - G\frac{\partial F}{\partial \nu} \right] d\sigma. \tag{7.18}$$

Indeed, the integral over Γ is a sum of the integral over Γ_1 and the integrals over Γ_i, $i = 2, \dots, N$. The last surface integrals over Γ_i, $i = 2, \dots, N$ all vanish since G and F satisfy the same boundary conditions. In the surface integral over Γ_1, we have, due to the boundary conditions, $G \equiv 0$ and $F \equiv 1$. Hence,

$$F(\mathbf{r}') = -D \int_{\Gamma_1} \frac{\partial G}{\partial \nu_1} \, d\sigma, \tag{7.19}$$

which completes the proof.

Now we give the following generalization of this reciprocity theorem.

Reciprocity Theorem. Assume that on all boundaries, including Γ_1, the Robin boundary conditions are prescribed:

$$-\frac{\partial G(\mathbf{r}, \mathbf{r}')}{\partial \nu_i} = S_i G(\mathbf{r}, \mathbf{r}') \quad \text{for} \quad \mathbf{r} \in \Gamma_i (i = 1, 2, \ldots, N). \tag{7.20}$$

The flux of particles to any boundary, say, Γ_k $(k = 1, 2 \ldots)$, defined as

$$I(\mathbf{r}) = -D \int_{\Gamma_k} \frac{\partial G(\xi, r)}{\partial \nu_k} \, d\sigma \tag{7.21}$$

satisfies the equality $I(\mathbf{r}) \equiv F(\mathbf{r})$ where $F(\mathbf{r})$ is the unique solution of the following homogeneous BVP:

$$D \Delta F(\mathbf{r}) - \tau^{-1} F(\mathbf{r}) = 0, \tag{7.22}$$

with the modified boundary condition

$$-\frac{\partial F(\mathbf{r})}{\partial \nu_k} = S_k (F(\mathbf{r}) - 1) \quad for \quad \mathbf{r} \in \Gamma_k \tag{7.23}$$

but with the same boundary conditions on other surfaces:

$$-\frac{\partial F(\mathbf{r})}{\partial \nu_i} = S_i F(\mathbf{r}) \quad \text{for} \quad \mathbf{r} \in \Gamma_i, \ i \neq k. \tag{7.24}$$

Proof. The proof follows by substitution of equality (7.23) in Green formula (7.18). Note that the case of the Dirichlet boundary conditions is a special case of this statement when $S_1 = \infty$.

This theorem can be used in many practically interesting cases when there is a need to calculate the particles fluxes on some specific parts of the boundary. In what follows, we show some practically interesting applications of this theorem. Let us start with a 1D case, which is simple. However, by using the splitting probabilities, quite a difficult problem of the nanowire growth simulation can be solved in the framework of this model.

7.4 Splitting and survival probabilities

7.4.1 Splitting probabilities for a finite interval and nanowire growth simulation

The epitaxial growth of nanowires is mainly controlled by the diffusion of Ga and N atoms over the surface of nanowires. In [100], we suggested a stochastic model of nanowire growth that includes some ad hoc parameters. Here, we suggest a direct simulation of the surface diffusion, and this makes the model self-consistent. The technique presented provides an independent algorithm for simulation of the growth of nanowires.

Let us find the splitting probabilities for a finite interval $[0, R]$. The governing diffusion equation reads

$$D\Delta n(x') - \frac{1}{\tau} n(x') = -\delta(x' - x), \quad x', x \in (0, L), \tag{7.25}$$

which is solved, generally, subject to the Robin boundary conditions:

$$-D\frac{\partial n}{\partial x} + S_1 n = 0, \quad \text{for} \quad x = 0,$$
$$D\frac{\partial n}{\partial x} + S_2 n = 0, \quad \text{for} \quad x = R.$$

In (7.25), τ is the mean diffusion time, and Δ stands for the 1D Laplace operator, $\Delta = \frac{d^2}{dx^2}$. To simplify the notations, throughout the chapter, we use the Δ-operator notation in any dimension and in any coordinate system.

According to the reciprocity theorem, the flux of particles to one of the boundaries (in our case, to one of the ends of the interval) which equals the splitting probability P_1 defined as the probability for a particle starting at x to hit the origin $x = 0$ before it reaches the right end of the interval $x = R$, satisfies the following BVP:

$$\Delta n(x) - \lambda^2 n(x) = 0, \qquad x \in (0, R), \tag{7.26}$$

$$-\frac{\partial n}{\partial x} + \frac{S_1}{D} n = \frac{S_1}{D}, \qquad \text{for} \quad x = 0, \tag{7.27}$$

$$\frac{\partial n}{\partial x} + \frac{S_2}{D} n = 0, \qquad \text{for} \quad x = R, \tag{7.28}$$

where $\lambda = \sqrt{1/D\tau}$ is the inverse diffusion length.

Let us first consider the case of no desorption, i.e., $\tau = \infty$. The general solution is $n = a + bx$ where the constants a and b are determined from the boundary conditions. This gives the explicit expression for the splitting probability P_1:

$$P_1 = \frac{S_1 D + S_1 S_2 (R - x)}{D(S_1 + S_2) + S_1 S_2 R}.$$

In the case when the particle is absorbed on both ends of the interval, i.e., when $S_1 = \infty$, $S_2 = \infty$, the probability P_1 has the following very simple form: $P_1 = 1 - x/R$. The probability to first reach the right end $x = R$ is thus $P_2 = x/R$.

In the general case when τ is finite, i.e., $\lambda > 0$, the general solution has the form

$$n(x) = a e^{\lambda x} + b e^{-\lambda x}.$$

The constants a and b are found from boundary conditions (7.27), (7.28). From these simple equations, we find the solution

$$n(x) = \frac{S_1}{D(\frac{S_2}{D} - \lambda)} \left[\frac{(\frac{S_2}{D} - \lambda)e^{\lambda x} - e^{-\lambda x} e^{2\lambda R}(\frac{S_2}{D} + \lambda)}{(\frac{S_1}{D} - \lambda)(\frac{S_2}{D} - \lambda) - e^{2\lambda R}(\frac{S_2}{D} + \lambda)(\frac{S_1}{D} + \lambda)} \right], \tag{7.29}$$

which is the probability P_1 that a particle starting at a point $x \in (0, R)$ first hits the left end $x = 0$ before it reaches the right end $x = R$ or is absorbed inside the interval. One important example is the case of absorption on both ends of the interval, i.e., when $S_1 = S_2 = \infty$. From general formula (7.29), we obtain the splitting probabilities in the form

$$P_{left} = \frac{\sinh[\lambda(R - x)]}{\sinh[\lambda R]}, \tag{7.30}$$

the probability to reach the left end $x = 0$, and

$$P_{right} = \frac{\sinh[\lambda x]}{\sinh[\lambda R]}, \tag{7.31}$$

the probability to reach the right end $x = R$.

In the case of a half-line, i.e., for $R = \infty$, we should take the solution in the form $n(x) = be^{-\lambda x}$ to satisfy the condition $n(\infty) = 0$. The survival probability in this case is obtained from the boundary condition at $x = 0$ and takes the form

$$n(x) = \frac{1}{1 + \frac{\lambda D}{S_1}} e^{-\lambda x}, \tag{7.32}$$

where x is the distance from the starting position to the origin $x = 0$.

Using the explicit representations for the splitting and survival probabilities, the simulation of a particle motion on the surface of a nanowire (considered as an interval) is simple. The diffusion on the nanowire surface of the incoming flux of Ga and N atoms is simulated as follows. A particle launched on the side surface of a nanowire at a point $x_1 \in (0, R)$ with the probability P_{left} hits the bottom of the nanowire, and with the probability P_{right}, it reaches the top of the nanowire where it makes the contribution to the nanowire growth. With probability $P_{desorb} = 1 - P_{left} - P_{right}$, the particle is desorbed and is randomly (with isotropic direction) scattered and hits one of the neighbouring nanowires at some other height at a point $x_2 \in (0, R)$. The trajectory is simulated till the particle reaches the top of a nanowire, or it flies out of the simulation region.

This technique avoids simulation of the diffusion trajectories of a particle and is therefore extremely fast.

7.4.2 Survival probability for a disc and the exterior of a circular cylinder

Let us start with the diffusion problem for the interior of a disk of radius R

$$D\Delta n(\mathbf{r}') - \frac{1}{\tau} n(\mathbf{r}') = \delta(\mathbf{r}' - \mathbf{r})$$

with the boundary condition

$$D\frac{\partial n}{\partial v}(R) + Sn(R) = 0.$$

The reciprocity theorem says that the survival probability satisfies the same equation but with zero source, and in the right-hand side of the boundary conditions, the zero term should be replaced with S/D.

The general solution of the homogeneous diffusion equation, which is rotationally symmetric, has the form

$$n(\mathbf{r}) = AI_0(r/L) + BK_0(r/L),$$

where we denote the diffusion length by $L = \sqrt{\tau D}$, and I_0, K_0 are the modified Bessel functions of the second kind. At $r = 0$, the solution must be finite, so $B = 0$, and the boundary condition reads

$$A\left[\frac{1}{L} I_1(R/L) + \frac{S}{D} I_0(R/L)\right] = \frac{S}{D},$$

which implies that the survival probability is given by

$$n = \frac{I_0(r/L)}{I_0(R/L) + \frac{D}{SL} I_1(R/L)}.$$

Consider now an infinite circular cylinder of radius R and the axis coinciding with the axis OZ of the cylindrical coordinate system. The diffusion of particles starting at $\mathbf{r}$ in the exterior of the cylinder is governed by the diffusion equation

$$D\Delta n(\mathbf{r}') - \frac{1}{\tau} n(\mathbf{r}') = \delta(\mathbf{r}' - \mathbf{r})$$

with the boundary condition

$$D\frac{\partial n}{\partial v}(R) + Sn(R) = 0.$$

In this case, we should satisfy the conditions at the infinity. Therefore, in the general solution $u(\mathbf{r}) = AI_0(r/L) + BK_0(r/L)$, we should take $A = 0$, since $I_0(\infty) = \infty$. Again, using the reciprocity theorem, we write down the Robin boundary conditions

$$B\mu K_1(\mu R) + B\lambda K_0(\mu R) = \lambda,$$

where $\lambda = S/D$, $\mu = 1/L$.

Thus, the survival probability in the exterior of the cylinder has the form

$$SProb = \frac{K_0(r/L)}{K_0(R/L) + \frac{D}{SL}K_1(R/L)}, \tag{7.33}$$

where $r = |\mathbf{r}|$. Therefore, the part of particles absorbed in the exterior of the cylinder reads

$$I(r) = 1 - SProb.$$

Note that for small values of r (say, $r < L/4$), the following approximation can be used:

$$SProb = \frac{\ln(r/2L) + \gamma}{\ln(R/2L) + \gamma - D/(S \cdot R)},$$

where $\gamma = 0.5772\ldots$ is the Euler constant.

The exact expression for the survival probability outside of an infinite cylinder will be used in the next section to validate the random walk algorithm for the cathodoluminescence imaging technique.

7.4.3 Splitting probabilities for concentric spheres and annulus

So far we have considered linear problems. Nonlinear problems where the diffusion particles are interacting with each other can also be solved by stochastic simulation algorithms, but probabilistic interpretations are not so easy to find. In the literature, the most known case concerns the homogeneous Smoluchowski coagulation equation where the pairwise interactions are simulated by the kinetic Monte Carlo (KMC) method. Inhomogeneous coagulation equations with diffusion were solved by some approximative models, which are very time consuming. In this chapter we suggest a very simple stochastic algorithm for solving this kind of nonlinear equations with diffusion, based on splitting probabilities for concentric spheres (3D) and annulus (2D).

Let us consider the following problem. A diffusing particle starts at a point $\mathbf{r}$ in an annulus G with the inner radius R_1 and outer radius R_2. Thus, $R_1 < r < R_2$, where $r = |\mathbf{r}|$. We are interested in splitting probabilities, i.e., in the probability P_1 that the particle reaches first the inner boundary without hitting the outer boundary, and in $P_2 = 1 - P_1$, the probability that the diffusing particle reaches first the outer boundary without hitting the inner boundary. No absorption inside the annulus is assumed. We assume general Robin boundary conditions. By the reciprocity theorem, the splitting probability, say, $P_1 = F(r)$, which is the flux to the inner boundary, satisfies the following BVP:

$$\begin{aligned} D\Delta F(r) &= 0, && r \in G, \\ D\frac{\partial F(r)}{\partial \mathbf{n}} + S_1 F(r) &= S_1, && \text{for } r = R_1, \\ D\frac{\partial F(r)}{\partial \mathbf{n}} + S_2 F(r) &= 0, && \text{for } r = R_2, \end{aligned} \tag{7.34}$$

where Δ is the radial Laplacian in 2D and $\mathbf{n}$ is the exterior unit normal vector to the boundary of the annulus.

Solution of this problem can be found explicitly. Indeed, the general solution that depends only on the radial variable r has the form

$$F(r) = A + B\ln(r),$$

where the constants are defined from the boundary conditions. These conditions read

$$\begin{aligned} -\frac{B}{R_1} + \frac{S_1}{D}[A + B\ln(R_1)] &= \frac{S_1}{D}, \\ \frac{B}{R_2} + \frac{S_2}{D}[A + B\ln(R_2)] &= 0. \end{aligned}$$

From these equations, we find the constants A and B, and the final expression for the probability $P_1 = F(r)$ is written in the form

$$F(r) = \frac{\ln\frac{R_2}{r} + \frac{D}{S_2 R_2}}{\ln\frac{R_2}{R_1} + \frac{D}{S_1 R_1} + \frac{D}{S_2 R_2}}. \tag{7.35}$$

Note that in the case of pure absorption on the inner and outer boundaries ($S_1 = \infty$, $S_2 = \infty$), this expression simplifies to

$$F(r) = \frac{\ln\frac{R_2}{r}}{\ln\frac{R_2}{R_1}}, \tag{7.36}$$

which coincides with the known result given in Redner's book [90].

Generalization to the 3D case, for a domain that is bounded by the inner sphere of radius R_1 and outer sphere of radius R_2, is straightforward, keeping in mind that the general spherically symmetric solution of the Laplace equation has the form $F(r) = A + B/r$. From the boundary conditions, we find the constants A and B, which gives the explicit expression:

$$F(r) = \frac{\left(\frac{R_2}{r} - 1\right) + \frac{D}{S_2 R_2}}{\left(\frac{R_2}{R_1} - 1\right) + \frac{D}{S_2 R_2}\left(1 + \frac{R_2^2}{R_1^2}\frac{S_2}{S_1}\right)}. \tag{7.37}$$

From this, in the case of pure absorptions on the boundaries, we find

$$F(r) = \frac{\frac{R_2}{r} - 1}{\frac{R_2}{R_1} - 1}, \tag{7.38}$$

which coincides with the result presented in [90].

In a more general case, when the diffusion equation for an annulus involves an absorption inside, written as $D\Delta F - \frac{1}{\tau}F = 0$, the general solution has the form

$$u(\mathbf{r}) = AI_0(r/L) + BK_0(r/L).$$

Then, in contrast to the case of a disc, both functions I_0 and K_0 are present, and the Robin boundary conditions on the inner and outer boundaries read

$$AF_1(R_1) + BF_2(R_1) = \frac{S_1}{D},$$
$$AF_3(R_2) + BF_4(R_2) = 0,$$

where

$$F_1(R_1) = \frac{S_1}{D}I_0\left(\frac{R_1}{L}\right) - \frac{1}{L}I_1\left(\frac{R_1}{L}\right), \quad F_2(R_1) = \frac{S_1}{D}K_0\left(\frac{R_1}{L}\right) + \frac{1}{L}K_1\left(\frac{R_1}{L}\right),$$
$$F_3(R_2) = \frac{S_2}{D}I_0\left(\frac{R_2}{L}\right) + \frac{1}{L}I_1\left(\frac{R_2}{L}\right), \quad F_4(R_2) = \frac{S_2}{D}K_0\left(\frac{R_2}{L}\right) - \frac{1}{L}K_1\left(\frac{R_2}{L}\right).$$

From these equations, we find the constants A and B; thus, the solution is

$$P_1 = \frac{S_1}{D}\frac{K_0(r/L) - \frac{F_4(R_2)}{F_3(R_2)}I_0(r/L)}{F_2(R_1) - \frac{F_4(R_2)}{F_3(R_2)}F_1(R_1)}.$$

Generalization to the case of a spherical shell between the outer sphere of radius R_2 and inner radius R_1, with general Robin boundary conditions, can be derived using the general solution

$$n(r) = a\frac{\sinh(\lambda r)}{r} + b\frac{\cosh(\lambda r)}{r}. \tag{7.39}$$

From the boundary conditions, the constants a and b are obtained, and we present the final results:

$$n(r) = \frac{\frac{S_1}{D}\left\{\frac{\sinh\lambda r}{r} - \frac{F_{21}(R_2)}{F_{22}(R_2)}\frac{\cosh\lambda r}{r}\right\}}{F_{11}(R_1) - \frac{F_{21}(R_2)}{F_{22}(R_2)}F_{12}(R_1)}, \tag{7.40}$$

where

$$F_{11}(R_1) = \frac{\sinh\lambda R_1}{R_1^2} + \frac{S_1}{D}\frac{\sinh\lambda R_1}{R_1} - \frac{\lambda\cosh\lambda R_1}{R_1},$$
$$F_{12}(R_1) = \frac{\cosh\lambda R_1}{R_1^2} + \frac{S_1}{D}\frac{\cosh\lambda R_1}{R_1} - \frac{\lambda\sinh\lambda R_1}{R_1},$$
$$F_{21}(R_2) = \frac{\lambda\cosh\lambda R_2}{R_2} + \frac{S_2}{D}\frac{\sinh\lambda R_2}{R_2} - \frac{\sinh\lambda R_2}{R_2^2},$$
$$F_{22}(R_2) = \frac{\lambda\sinh\lambda R_2}{R_2} + \frac{S_2}{D}\frac{\cosh\lambda R_2}{R_2} - \frac{\cosh\lambda R_2}{R_2^2}.$$

The case for the exterior of a sphere, when $R_2 \to \infty$, is treated the same way, but we have to take $b = 0$ in general solution (7.39). The solution that is the survival probability for a particle starting at a distance $h = r - R$ from the surface of the sphere of radius R has the form

$$n(r) = \frac{R}{r}\frac{e^{-\lambda h}}{1 + \frac{\lambda D}{S} + \frac{D}{SR}}. \tag{7.41}$$

Remarkable, this result converges, as $R \to \infty$, to the relevant 1D result for a half-line given by formula (7.32). This confirms that the survival probability of a 3D diffusion process for a half-space is defined by its 1D z-component only since $R/(R+h) \to 1$ as $R \to \infty$.

In the problem of the cathodoluminescence, we will need the survival probability for a particle inside a sphere. The diffusion equation reads $\Delta n - \lambda^2 n = 0$ with the absorption boundary condition $n(R) = 1$ as follows from the reciprocity theorem. The general spherical symmetric solution of this diffusion equation is given by (7.39). In our case, $b = 0$, since the solution is finite at the origin. The boundary conditions yield $a = \frac{R}{\sinh(\lambda R)}$; thus, we conclude that the survival probability is given by

$$P_{surv} = n(r) = \frac{\sinh(\lambda r)}{r}\frac{R}{\sinh(\lambda R)}. \tag{7.42}$$

In the cathodoluminescence problem, we will use the probability $P_{absorpt} = 1 - P_{surv}$, which is the probability that the particle starting inside the sphere at the distance r from the centre will be absorbed inside the sphere before it hits the surface of the sphere.

We will need also the survival probability for a hemisphere with the same absorption boundary conditions on the spherical surface, and with the reflection (Neumann) boundary conditions on the plane part of the boundary. Due to the symmetry of the Wiener process with respect to the plane part of the boundary, it is clear that the survival probability has the same form given by (7.42).

7.5 Cathodoluminescence

The cathodoluminescence imaging is a relatively new technique, which is used for imaging of microstructures like the threading dislocations in semiconductors (e.g. see [77, 82]), studies of sedimentary rocks [6] and many others. In this section, we apply the presented stochastic algorithm to the imaging of dislocations, which are considered as circular half-cylinders in a half-space, with the boundary $z = 0$.

So let us consider an unbounded domain G, which is defined as the half-cylinder's exterior overlapped with the half-space $z < 0$, see the general scheme presented in Fig. 7.1. We denote the plane $z = 0$ by Γ_1 and the cylinder's boundary by Γ_2. It is assumed that the cylinder axis coincides with the coordinate axis z.

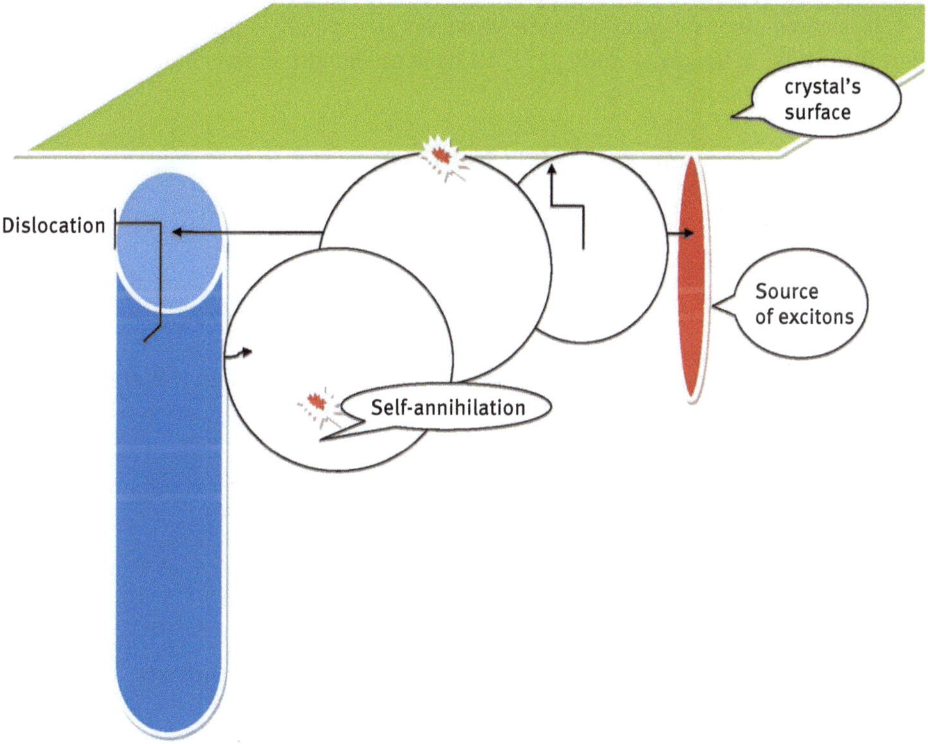

Fig. 7.1. Random walk scheme for the cathodoluminescence imaging.

We assume that a source of excitons $I(\mathbf{x})$ is placed in the domain G. The exciton is diffusing in the domain G with a constant diffusion coefficient D. When reaching the plane boundary $z = 0$, the exciton can be absorbed with a rate proportional to S_1. If not absorbed, it continues to diffuse in G. The same for the surface of the cylinder: when reaching Γ_2, the exciton is absorbed with a rate proportional to S_2. It means that both on the plane boundary and on the side surface of the cylinder, the Robin boundary conditions are prescribed. The exciton can be self-annihilated in the domain G, and the mean lifetime of an exciton before it annihilates in G is τ.

The concentration density of excitons at a point x in G is governed by the following equation:

$$D\Delta n(\mathbf{x}) - \frac{1}{\tau} n(\mathbf{x}) + I(\mathbf{x}) = 0, \qquad \mathbf{x} \in G \tag{7.43}$$

and satisfies the boundary conditions:

$$D\frac{\partial n(\mathbf{x})}{\partial \nu} + S_1 n(\mathbf{x}) = 0, \qquad \mathbf{x} \in \Gamma_1, \tag{7.44}$$

$$D\frac{\partial n(\mathbf{x})}{\partial \nu} + S_2 n(\mathbf{x}) = 0, \qquad \mathbf{x} \in \Gamma_2. \tag{7.45}$$

Here, ν is the external unit normal vector at the point $\mathbf{x}$ on the boundary.

It is convenient to introduce new variables $l_D^2 = \tau D$, $l_{S_1} = \tau S_1$ and $l_{S_2} = \tau S_2$, so that the diffusion length l_D and l_{S_1}, l_{S_2} all have the length dimensionality. Equations (7.43)–(7.45) then read

$$\Delta n(\mathbf{x}) - \frac{1}{l_D^2} n(\mathbf{x}) + I(\mathbf{x})/D = 0, \qquad \mathbf{x} \in G, \tag{7.46}$$

$$\frac{\partial n(\mathbf{x})}{\partial \nu} + \frac{l_{S_1}}{l_D^2} n(\mathbf{x}) = 0, \qquad \mathbf{x} \in \Gamma_1, \tag{7.47}$$

$$\frac{\partial n(\mathbf{x})}{\partial \nu} + \frac{l_{S_2}}{l_D^2} = 0, \qquad \mathbf{x} \in \Gamma_2. \tag{7.48}$$

We are interested in calculation of the total concentration of self-annihilated excitons in our domain G.

To this end, we integrate equation (7.46) over the domain G:

$$\int_G \Delta n(\mathbf{x}) d\mathbf{x} - \frac{1}{l_D^2} \int_G n(\mathbf{x})\, d\mathbf{x} + \int_G I(\mathbf{x})\, d\mathbf{x} = 0. \tag{7.49}$$

Using the Green formula and the boundary conditions, we find

$$-\frac{l_{S_1}}{l_D^2} \int_{\Gamma_1} n(\mathbf{x}) d\mathbf{x} - \frac{l_{S_2}}{l_D^2} \int_{\Gamma_2} n(\mathbf{x}) d\mathbf{x} - \frac{1}{l_D^2} \int_G n(\mathbf{x})\, d\mathbf{x} + \int_G I(\mathbf{x})\, d\mathbf{x} = 0. \tag{7.50}$$

This shows that all the excitons originated from the source are distributed over three parts: the first is the set of excitons absorbed on the boundary Γ_1, the second are those absorbed on Γ_2 and the third consists of excitons self-annihilated in G. Our goal is to calculate the concentration of self-annihilated excitons relative to the total concentration of excitons originated from the source.

7.5.1 The random WOS and hemispheres algorithm

The random WOS algorithm [98] for solving the problem is based on the probabilistic interpretation of the diffusion process and the strict Markov property.

The general idea behind the method is in tracking the excitons in our domain, and calculate the number of excitons that were self-annihilated in G. We can do it without introducing any grid. Indeed, let us consider a sphere with the centre placed at the exciton position generated by the source, and the radius r equal to the minimal distance to the boundary. It is known (e.g. see [98]) that the survival probability of the diffusion process inside this sphere is given by $Pr = l_D\, r / \sinh(l_D\, r)$, while the random exit point from this sphere is uniformly distributed on the surface of the sphere. Based on this information, we can track the exciton trajectory: if the exciton survives in the first sphere, we consider the exit point as the centre of the second sphere. This process is then repeated until the exciton reaches the boundary or annihilates inside a sphere. When reaching the boundary, say, Γ_1, the exciton is absorbed with probability $\frac{l_{\Gamma_1} h/l_D^2}{(1+l_{\Gamma_1} h/l_D^2)}$, otherwise the exciton is reflected to the distance h along the normal vector inside the domain G and proceeds to make the next diffusion step. Here, h is small enough.

For the plane boundary with the reflection (Neumann) boundary conditions, the random walk simulation can be made more efficient. The case when on the plane boundary the reflection (Neumann) boundary conditions are imposed, the standard described random WOS algorithm can be improved. Indeed, in the standard implementation, the random walk may have long excursions near the boundary. To avoid multiple recurrence and long trajectories, we choose hemispheres instead of spheres, with the plane part coinciding with the plane boundary. So assume we are at a point whose distance to the plane boundary is less than to the cylinder's surface. Then, we construct a hemisphere with this point lying on a radius that is perpendicular to the plane boundary. Now we have to find the survival probability of a Brownian motion inside this hemisphere, under the condition that on the plane boundary it is reflected, while on the spherical surface it is absorbed. Due to the symmetry of the reflecting Wiener process with respect to the plane, we conclude that the survival probability of the Brownian motion in the hemisphere with this boundary conditions is equivalent to the survival probability in a sphere with the same radius but with absorption boundary condition on the whole sphere. This survival probability is equal to $\lambda R/\sinh(\lambda R)$ provided the particle starts at the centre of the sphere. However, in our case, the particle inside the sphere has an arbitrary position. For this case, we have derived above the explicit expression for the survival probability (7.42).

It remains to find the exit point of the reflected Brownian motion in the hemisphere. As explained, due to symmetry, we need to simulate the exit point of a particle starting at an arbitrary point in the hemisphere, and in the case the exit point x^* is on the other side of the plane ($z > 0$), we just take a point inside the sphere that is symmetric to the point x^*.

Simulation of the exit point in a sphere that starts at an arbitrary point r inside the sphere can be efficiently simulated as follows.

First, in the point r, an isotropically distributed unit vector is simulated, which defines a line along this direction. This line intersects the sphere in two points, say, point A and point B. Let the distance from r to A be d_A and the distance from r to B be d_B. Then, the exit point is taken as the point A with probability $d_B/(d_A + d_B)$, and the exit point is taken as B with probability $1 - d_B/(d_A + d_B)$.

To justify this algorithm, we mention that (1) a normal projection of the Wiener process inside the sphere on any line is a 1D Wiener process on this line, and (2) the exit probabilities of a Wiener process on a segment [A,B] starting at a point $r \in [A, B]$ to the ends of the segment are given by the quantities $d_B/(d_A + d_B)$ and $d_A/(d_A + d_B)$.

Let us prove these two properties. First, concerning the projection. Let us consider a 3D Wiener process w_3 starting at a point x_0. The orthogonal projection of this process on any line that goes through this point is also a Wiener process on this line. Let us denote the unit vector that defines the line by n. The orthogonal projection of w_3 on the line is defined as a linear transformation of w_3:

$$w_1 = \frac{(w_3 \cdot n)}{(n \cdot n)} n.$$

Thus, w_1 is obtained from w_3 by this linear transform. Therefore, w_1 is also Gaussian, it has also independent homogeneous increments and they both start from the same point. Note that a much more complicated proof of this simulation technique was given in [45].

Summarizing, let us describe the algorithm based on the random walk on spheres and hemispheres. Generally, the domain where we consider the transport of diffusing excitons can be considered as a half-space overlapped with a set of non-overlapping circular cylinders whose axes are perpendicular to the plane $z = 0$, the boundary of the domain. This boundary is denoted by Γ_1, and the set of all side boundaries of the cylinders is denoted by Γ_2. Thus, our domain has the plane boundary Γ_1 and the boundary Γ_2. The governing equations and the Robin boundary conditions are given by (7.43–7.45). The algorithm can be described by the following steps.

(1) Generate the starting position r_0 according to the given source $I(r)$.
(2) Find the minimal distance, $d_{\min}$, from r_0 to the boundary of the domain. It may be the distance d_{Γ_1} to the plane Γ_1 or d_{Γ_2}, the distance to the surface of one of the cylinders.
(3) If $d_{\Gamma_1} < d_{\Gamma_2}$, i.e., the particle is closer to the plane boundary, we construct a hemisphere of radius d_{Γ_2} (calculated as the distance to the closest cylinder) such that the particle lies on the radius that is perpendicular to the plane boundary. Calculate the survival probability

$$Ps = \frac{\sinh(\lambda r_0)/\lambda r_0}{\sinh(\lambda R)/\lambda R},$$

where $R = d_{\Gamma_2}$.
(4) With the probability $1 - Ps$, the trajectory is terminated, and with probability Ps, we make the next step: simulate the exit point from the hemisphere. To this end, we simulate, as described above, the isotropic direction at the point x_0 and find the points of intersection of the line having this direction with the sphere that is obtained by the symmetrical extension of the hemisphere relative to the plane boundary. If the simulated point x_1 belongs to the hemisphere, it implies that we have found the exit point, so we put $x_0 = x_1$ and go to step 2. If the point is outside of the domain (i.e., it is lying on the reflected hemisphere), we just reflect this point symmetrically with respect to the plane and thus get the point x_1. Next, we put $x_0 = x_1$ and go to step 2.
(5) If $d_{\Gamma_1} > d_{\Gamma_2}$ in step 3, then we construct the sphere of radius d_{Γ_2} and calculate the survival probability $Ps = \lambda d_{\Gamma_2}/\sinh(\lambda d_{\Gamma_2})$. Now, with the probability $1 - Ps$, the trajectory is finished, and with probability Ps, we sample the exit point on the sphere that is a point uniformly distributed on this sphere.
(6) If on some step, say, k, the walking particle is so close to the boundary Γ_2 that the distance from the point to the surface Γ_2 is less than a fixed small value ε, then the trajectory stops, if we deal with the absorption case ($S_2 = \infty$). If two boundary events are possible, absorption or reflection, we proceed with sampling as described above at the beginning of section 7.5.1, taking the jump step $h \sim \sqrt{\varepsilon}$.
(7) The desired signal is calculated as $F = M/N$, where M is the number of trajectories absorbed inside the domain and N is the total number of simulated trajectories.
(8) The steps 1–7 are repeated for a set of starting points $\{x_k\}$, and the desired signal is evaluated as $F(x_k)$, a function of the distance from the dislocation to the point source.

It should be noted that the simulated trajectory of excitons always tends to reach the boundary. Strictly speaking, the Markov process of randomly walking excitons converges to the boundary; however, it reaches the boundary in infinite number of steps. To make the number of steps finite, one introduces an ε-boundary (ε-strip) in the neighbourhood of the boundary Γ_2, as mentioned implicitly in the description of the algorithm in Step 6. This makes the number of steps in the random walk process finite, with a remarkably small mean value: $\langle N\rangle \sim |\log(\varepsilon)|$. In simulations, we have taken $\varepsilon = 10^{-4}$. The error due to the introduction of ε-boundary is of order of ε, so it is negligible compared to the experimental errors.

We are interested in calculation of the function $F(x)$, the total number of self-annihilated excitons in G, divided by the total number of excitons originated from the source, placed at the distance x from Γ_2, the cylinder's boundary.

To validate the algorithm, we compared the simulations with the exact result given by formula (7.33) for the case when the domain is the exterior of an infinite cylinder, and obtained a perfect agreement. We show now the simulation results for a more general case, for the domain composed of one semi-infinite circular cylinder and the half-space described above. In this case, the contrast curve can be well approximated analytically using the probabilistic interpretation and the survival probability.

The 3D diffusion can be considered as a superposition of the radial diffusion and independent 1D diffusion along the axis OZ. The survival probability of the radial component is given by (7.33), and the 1D diffusion can be treated as a diffusion on the ray $[0, -\infty)$. The survival probability of a particle starting at a point $-z$ with the Robin boundary condition with the surface recombination coefficient S_1 was derived above (see (7.32) and (7.41)); it reads as

$$P_1 = \frac{1}{1 + \frac{D}{S_1 L}} \mathrm{e}^{-|z|/L}.$$

In a small neighbourhood around the circle, which is the intersection of the cylinder and the half-space, the 1D and the radial diffusions interact strongly, which however cannot affect much the integrals of the solution in the whole domain.

To get closer, let us write down the solution n in the form $n(r,z) = n_1(z) + n_2(r)$, where both $n_1(z)$ and $n_2(r)$ solve the same diffusion equation. These two solutions are coupled by the boundary conditions. The function $n_2(r)$ is the solution in the exterior of the cylinder, and the exact expression for this function is given by formula (7.33). For the function $n_1(z)$, we come to the problem

$$\begin{aligned} &D\frac{d^2 n_1(z)}{dz^2} + S_1 n_1(z) = 0, \\ &\frac{dn_1}{dz} + \frac{S_1}{D}(n_1 + n_2) = \frac{S_1}{D}, \qquad z \in \Gamma_1 \end{aligned}$$

since $\frac{dn_2}{dz} = 0$. We use the general form of solution, $n_1 = a\mathrm{e}^{-z/L}$, in the boundary conditions and find the constant a; thus,

$$P_1 = n_1 = \frac{1 - u_2(r)}{1 + \frac{D}{S_1 L}} \mathrm{e}^{-z/L}.$$

Note that we have made only the first-order iteration, assuming that on the boundary $z = 0$, we use the solution $n_2(r)$ for the exterior of the infinite cylinder. In the second iteration, we should take into account the value $n_1(z)$ on the boundary $r = R$, etc. This leads to a correction factor β in the formula for P_1, which we have analyzed numerically

$$P_1 = \frac{1 - u_2(r)}{1 + \frac{D}{\beta S_1 L}} e^{-z/L} .$$

Note that the factor β is close to the value $\beta = 8$, and it slowly varies with the change of the diffusion length L.

Therefore, using expression (7.33) for the function $n_2(r)$, we conclude that for the surface recombination rate equal to S_2 on the cylinder's side and equal to S_1 on the plane $z = 0$, the concentration of absorbed excitons in the whole domain is given by the following approximation:

$$I = 1 - n_1 - n_2 = \left[1 - \frac{K_0(r/L)}{K_0(R/L) + \frac{D}{S_2 L} K_1(R/L)}\right] \frac{1}{1 + \frac{\beta L S_1}{D}} , \tag{7.51}$$

where r is the radial coordinate of the starting point and $h = z$, where $-z$ is its z-coordinate; we have assumed here that $|z| << L$.

Let us give an example of the calculated function $F(x)$. We have fixed the radius of the semi-infinite cylinder as $R = 1$ nm. The source has been chosen as a finite cylinder of radius 1 nm and length 1 nm, with the axis parallel to the axis z. In Fig. 7.2, we show the behaviour of the function $F(x)$ for different values of S_1, for fixed value of the diffusion coefficient $D = 700000\, nm^2/ns$, $S_2 = \infty$ and $\tau = 1$. All simulation results

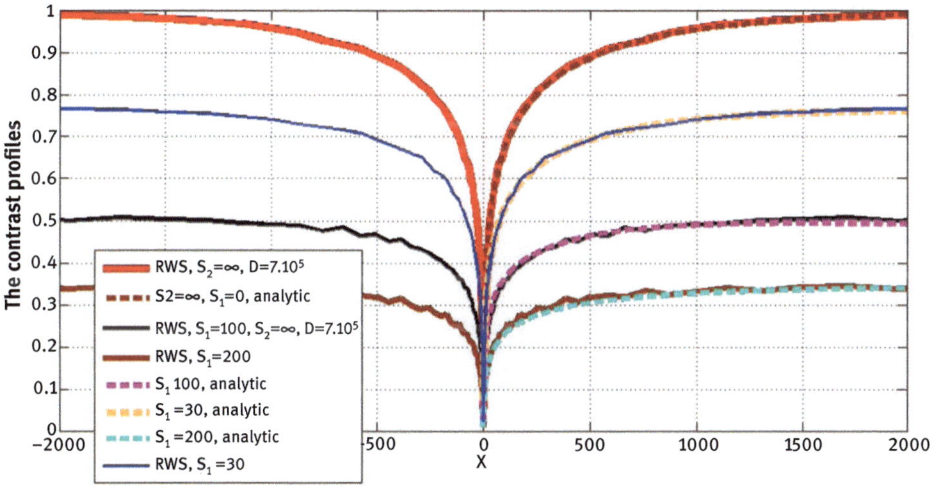

Fig. 7.2. The contrast profiles $n(r)$, for $D = 700000$, $S_2 = \infty$ and different values of S_1. Comparison of simulations with analytical approximation (7.51)

are compared against the analytical approximation given by (7.51). It is seen that the simulation results and the analytical approximation given by (7.51) are in perfect agreement.

7.6 Conclusion and discussion

Fast stochastic algorithms based on the exact representations for the first passage time and exit point probability densities, splitting and survival probabilities are developed. Three examples where the suggested methods can be efficiently applied are considered: (1) simulation of epitaxial nanowire growth, (2) diffusion imaging of microstructures, in particular, cathodoluminescence imaging for threading dislocations, and (3) simulation of the annihilation of electrons and holes in amorphous semiconductors in the vicinity of nonradiative centres, and quantum efficiency evaluation. The last example illustrates that the random WOS method can be used to solve nonlinear diffusion equations and more general classes of nonlinear Smoluchowski equations, if properly combined with the KMC method. We were able to obtain some analytical results for a practically interesting problem of estimating the cathodoluminescence signal for the case of a domain composed of a semi-infinite right circular cylinder and the half-space, which presents an excellent benchmark problem for the numerical methods validation.

8 A random WOS-based KMC method for electron–hole recombinations

To simulate spatially inhomogeneous bimolecular reactions driven by diffusion and tunnelling, we present here a new stochastic algorithm, which combines the random WOS method and the KMC algorithm. In the case of diffusion, and especially for low concentrations, such approach drastically decreases the computer simulation time. This is the case, for instance, when the annihilation of spatially separate electrons and holes in a disordered semiconductor is studied. The method treats all kinds of reactions involved in a unified stochastic kinetic scheme. In particular, along the diffusion and tunnelling, nonradiative recombinations in defect sites are taken into account. To validate the simulation algorithm, we compare our results with the asymptotics of the intensity of annihilation, which is known from theoretical predictions. Also, we compare the stochastic algorithms with finite-difference (FD) methods. More technical details can be found in our papers [51, 100] and [93].

8.1 Introduction

The fluctuation-limited kinetics of diffusion-controlled bimolecular reactions plays a crucial role in many physical, chemical and biological processes. In practice, however, the diffusion equations are often treated macroscopically, ignoring density fluctuations. In the absence of nonlinear interactions such as chemical reactions, the macroscopic diffusion equations govern uniform concentration distributions. Thermal fluctuations, initial density inhomogeneities and the randomness of reaction events lead to non-uniform concentration fields, thus drastically changing the solution, and in particular, the time dependence of the density for asymptotically long times.

Let us consider for simplicity a reaction of two types of particles, A and B, $A + B \to P$. The simplest kinetic approach to this reaction first considered by Smoluchowski [112] is based on the rate equations for concentrations

$$dn_A/dt = -g\,n_A(t)n_B(t), \quad dn_B/dt = -g\,n_A(t)n_B(t), \quad n_A(0) = a_0, n_B(0) = b_0,$$

where g is the reaction rate. For diffusion-controlled reactions in the dimension three, Smoluchowski obtained $g = 4\pi D r_0$, where r_0 is the particle radius and D is the relative diffusion coefficient. This equation can be solved explicitly: $n_A(t) = a_0/(1 + a_0 g t)$ for $a_0 = b_0$ and $n_A(t) = a_0 f_0/[b_0 \exp(f_0 g t) - a_0]$ for $a_0 < b_0$, where $f_0 = b_0 - a_0$.

The basic assumption underlying this reaction is a homogeneous spatial distribution of particles during the reaction at any time instant, i.e., the components A and B should be always perfectly mixed. Independent of the dimension d, this assumption results in solutions with a long-time asymptotics $\sim \exp(-f_0 g t)$ if $a_0 \neq b_0$ and $\sim 1/gt$ if $a_0 = b_0$.

Fluctuations lead to drastically different solutions. In particular, spatial correlations of particles may lead to segregation, i.e., the formation of spatially separated clusters composed entirely of particles of either type A or B. The formation of these clusters slows down the reaction dramatically, because only particles near the boundary of the clusters are likely to react, while particles inside the cluster have to diffuse to the boundary before they have a chance to react with a particle of the other type. In other words, fluctuations induce the formation of a mosaic of continuously growing domains, which contain only one of the two components, A or B. It was first shown by Ovchinnikov and Zeldovich [81] that in this case, the long-time asymptotics is $\sim t^{-3/4}$ if $a_0 = b_0$. Generally, for $a_0 = b_0$, the asymptotics $\sim t^{-d/4}$ is valid for any dimensionality $d \leq 4$, while for $d \geq 4$, $\sim t^{-1}$. This law was obtained by several authors using different arguments (see, e.g. [7] and [56]).

KMC methods are well developed for solving homogeneous Smoluchowski equations, which govern the coagulation (coalescence) of one type of particles of different sizes (e.g. see [102]). The method can be easily extended to particles of different types having different masses m_i (e.g. see [99]):

$$\frac{\partial n(m,t)}{\partial t} = \frac{1}{2}\int_0^{m_1}\cdots\int_0^{m_s} R(u, m-u)n(m-u,t)n(u,t)du$$
$$-n(m,t)\int_0^{\infty}\cdots\int_0^{\infty} R(u,m)n(u,t)du. \tag{8.1}$$

Here, m_i is the mass of the ith component in a particle, and m is a vector of compositions $(m_1, \ldots, m_s)$, where s is the total number of components; $n(m,t)dm$ is the number of particles having mass of component i in the range $[m_i, m_i + dm_i]$ at time t, and $R(u,m) = R(m,u)$ is the binary coagulation coefficient.

Extensions of the stochastic method to inhomogeneous Smoluchowski equation were suggested in our papers [53, 101]; however, the inhomogeneity was entered not through a diffusion but via a non-zero flow of the host gas. Diffusion was included in our next paper [101] where we extend the method using a stochastic Lagrangian trajectory–based algorithm. In all these studies, as well as in studies of other authors we are aware, the diffusion was modelled by small time steps of the particles. The only stochastic method where large time jumps were used has been studied in a series of papers by Opplestrup and coauthors (e.g. see [18, 79]). This method however suffers from the need to keep the evolution in time and space of all diffusing particles that complicates the algorithm and increases the computer time. A method that is close to this approach was used in our recent paper [100]. In this chapter, we present a new method that combines the well-known random WOS method and the KMC method for solving Smoluchowski coagulation equations. This method is extremely fast, especially for low concentrations, which is the case of the electron–hole annihilation process we study.

The numerical solution of the Smoluchowski-type equations discussed above is a highly challenging problem even for only a few particle types in the spatially inhomogeneous case. We deal in this chapter with three particle types (electrons, holes and nonradiative recombination centres), but the problem is inhomogeneous in space, with random distribution of the initial concentrations. This is the first and major difficulty of this problem. The second difficulty of our problem is caused by the low particle densities. Third, we are interested in the particle kinetics for very long times. Finally, we deal with stochastic initial conditions, so we have to take an average over the ensemble of initial distributions. Conventional numerical methods are not suitable for handling problems of this kind, and we therefore extend the Monte Carlo methods we have developed previously in [53, 102], and [100, 101] to the general case we consider in this chapter.

In this chapter, we deal with the recombination of electrons and holes in an inhomogeneous semiconductor. This topic has attracted considerable experimental and theoretical interest during the past five decades since the optoelectronic properties of technologically important materials have been found to be controlled by the electron–hole recombination dynamics. Recently, it has been discovered that (In,Ga)N/GaN quantum wells, used in light-emitting diodes for solid-state lighting, exhibit properties analogous to amorphous semiconductors due to the spatial localization of charge carriers and their recombination as spatially separate electron–hole pairs [8, 9, 37].

We consider that generally both electrons and holes are able to diffuse and recombine with each other either radiatively (i.e., giving rise to the generation of a photon) or nonradiatively via immobile recombination centres. Since electrons and holes are, in general, spatially separated, these recombination processes are considered to be induced by tunnelling from one localization site to the other and to be assisted by diffusion, allowing electrons and holes to meet each other in the same spatial location. What governs the recombination dynamics are both the fluctuations of the initial spatial distributions of electrons and holes and the fluctuations in their diffusivity resulting from the random depth of the localization sites. Hence, spatial correlations are crucial for the evaluation of the mean concentrations.

8.2 The mean field equations

We consider electrons, holes and nonradiative recombination centres distributed in a domain G with densities $n(\mathbf{r};t)$, $p(\mathbf{r};t)$ and $N(\mathbf{r})$, respectively. Some of these recombination centres with a density $N^+(\mathbf{r};t)$ are in the state waiting for capturing an electron, while the remaining centres $N(\mathbf{r}) - N^+(\mathbf{r};t)$ are waiting for a hole.

The radiative recombination of spatially separate electron–hole pairs is due to tunnelling with the rate $B(r) = B_0 \exp(-r/a)$, where r is the distance between an electron and a hole. The decrease of the electron density at the location $\mathbf{r}$ due to the recombination with holes in the vicinity is $-n(\mathbf{r};t)\int B(|\mathbf{x}|)p(\mathbf{r}+\mathbf{x};t)d\mathbf{x}$. The decrease of the electron density due to an interaction with a nonradiative recombination centre is described analogously: $-n(\mathbf{r};t)\int b_n(|\mathbf{x}|)N^+(\mathbf{r}+\mathbf{x};t)d\mathbf{x}$. Here $b_n(r) = b_{n0}\exp(-r/a_n)$, and the same kernel is defined for the hole-recombination centre interaction: $b_p(r) = b_{p0}\exp(-r/a_p)$.

The electron can also diffuse spatially with a diffusion coefficient $D_n(\mathbf{r})$ that may vary randomly with the spatial position $\mathbf{r}$. Altogether, the electron density thus follows the equation

$$\frac{\partial n(\mathbf{r};t)}{\partial t} = D_n(\mathbf{r})\Delta n(\mathbf{r};t) - n(\mathbf{r};t)\int B(|\mathbf{x}|)p(\mathbf{r}+\mathbf{x};t)d\mathbf{x} - n(\mathbf{r};t)\int b_n(|\mathbf{x}|)N^+(\mathbf{r}+\mathbf{x};t)d\mathbf{x}. \tag{8.2}$$

Analogously, the holes' density is described by the equation

$$\frac{\partial p(\mathbf{r};t)}{\partial t} = D_p(\mathbf{r})\Delta p(\mathbf{r};t) - p(\mathbf{r};t)\int B(|\mathbf{x}|)n(\mathbf{r}+\mathbf{x};t)d\mathbf{x} - p(\mathbf{r};t)\int b_p(|\mathbf{x}|)[N(\mathbf{r}+\mathbf{x}) - N^+(\mathbf{r}+\mathbf{x};t)]\,d\mathbf{x}. \tag{8.3}$$

The number of recombination centres waiting for an electron is reduced when the electron is captured and increased when the hole is captured:

$$\frac{\partial N^+(\mathbf{r};t)}{\partial t} = -n(\mathbf{r};t)\int b_n(|\mathbf{x}|)N^+(\mathbf{r}+\mathbf{x};t)d\mathbf{x} + p(\mathbf{r};t)\int b_p(|\mathbf{x}|)[N(\mathbf{r}+\mathbf{x}) - N^+(\mathbf{r}+\mathbf{x};t)]\,d\mathbf{x}. \tag{8.4}$$

Without loss of generality, we assume that the initial numbers of electrons and holes are equal (say, to n_0). At the initial time $t = 0$, the electrons, holes and recombination centres are randomly distributed:

$$n(\mathbf{r};0) = \sum_{i=1}^{n_0}\delta(\mathbf{r}-\mathbf{r}_i), \quad p(\mathbf{r};0) = \sum_{j=1}^{n_0}\delta(\mathbf{r}-\mathbf{r}_j), \quad N(\mathbf{r};0) = \sum_{k=1}^{N^{(0)}}\delta(\mathbf{r}-\mathbf{r}_k). \tag{8.5}$$

Here $\mathbf{r}_i$, $\mathbf{r}_j$ and $\mathbf{r}_k$ are independent uniformly distributed random positions, and $N^{(0)}$ is the total number of recombination centres. At $t = 0$, we can assume that all recombination centres are waiting for an electron, i.e., $N^+(\mathbf{r};0) = N(\mathbf{r};0)$ if the background doping density n_s is zero. Finally, the experimentally observed radiative intensity $I(t)$ is proportional to the photon flux $\phi(t)$ due to the generation of photons by the radiative recombination of electrons and holes:

$$\phi(t) = \left\langle \int \frac{1}{|G|} d\mathbf{r} \int B(|\mathbf{x}|) n(\mathbf{r}; t) p(\mathbf{r} + \mathbf{x}; t) d\mathbf{x} \right\rangle, \tag{8.6}$$

where the angle brackets stand for the mathematical expectation with respect to the initial random distribution of electrons, holes and recombination centres, and $|G|$ is the volume of the domain G. Since $I(t) = \hbar\omega\phi(t)$, where $\hbar\omega$ is the photon energy, we use the terms photon flux and intensity interchangeably, following [100].

Generally, in d dimensions, the long-time asymptotics in the case of purely radiative recombination is $n(t) \sim [\log(t)]^{-d}$ if only the electron–hole correlations are taken into account (see also [56]), i.e., for two dimensions,

$$\phi(t) \sim t^{-1} (\log t)^{-3}. \tag{8.7}$$

If, in addition, the electron–electron and hole–hole correlations are taken into account, then it leads to the following asymptotics (e.g. see [56, 100]), $n(t) \sim [\log(t)]^{-d/2}$. For the photon flux, this result implies $\phi(t) \sim t^{-1} [\log(t)]^{-(d/2+1)}$.

8.3 Monte Carlo Algorithms

The system of equations (8.2)–(8.4) has exactly the same structure as the inhomogeneous Smoluchowski coagulation equations (see, e.g. [101]) for three distinct species, in our case, the electrons, holes and recombination centres. The Smoluchowski equations may be interpreted probabilistically as an equation generated by Markov chains describing the pair interactions (e.g. see [53]). In [101], we developed a Monte Carlo algorithm, which solves inhomogeneous Smoluchowski equations with diffusion included. In this method, we turn to Lagrangian coordinates, and the solution was obtained by averaging over the ensemble of Lagrangian trajectories. In this elegant technique, however, the small time step evolution simulation was not avoided, so the method takes a lot of computer time, especially for the practically interesting case of low concentrations.

8.3.1 Random WOS for the diffusion simulation

The random WOS algorithm is known to be very efficient for solving some BVPs for elliptic and parabolic equations, see, for instance [98]. This approach is used here to accelerate the KMC algorithm for solving inhomogeneous Smoluchowski-type equations with diffusion.

The main idea behind the method we present here can be described as follows. For simplicity, we consider the case of mobile electrons and immobile holes and recombination centres. We treat the reactions due to diffusion and coagulation (in our case, annihilation) in a unified scheme. However, it cannot be done directly since the frequencies of possible interactions due to diffusion would be much higher compared to annihilations, since we have to keep the time step very small to correctly simulate the diffusion. Therefore, we introduce a large time step of diffusion as follows. First, we place spheres (or circles, in 2D) around each diffusing particle (electron, in our case) so that the centre is placed at the position of the particle, and the radius of each sphere is equal to the distance to the closest immobile particle (a hole or a nonradiative centre, in our case). Inside these spheres, the electrons are diffusing independently of each other, at least during the time τ_{min}, the minimum of $\tau_1, \ldots, \tau_n$, the random exit times of the diffusing particles from these spheres. The relevant frequencies of the events to reach the boundaries can be defined as

$$\lambda_i = \frac{1/\tau_i}{\sum_{j=1}^{n} 1/\tau_j}. \tag{8.8}$$

After this introduction of frequencies, there is no difference in the type of interactions due to diffusion and tunnelling; they differ only in the time scales, which are now comparable. So to apply the KMC method, we need only to calculate the frequencies of all the events in the system.

Note that the separation spheres can be placed quite arbitrarily, depending on a specific configuration of the electrons, holes and recombination centres. For instance, the centres of the spheres may not coincide with the electron positions. The only quantities we need to simulate are the random exit time from the sphere and the exit point on the sphere. When the particle starts from the centre, the exit point is obviously uniformly distributed on the sphere. As for the random exit time, we need only to simulate it for a sphere of the unit radius since for the Wiener process (in any dimension), $\tau_R = \tau_1 R^2/2D$, where τ_R is the random exit time for a sphere of radius R, τ_1 is the exit time of a standard Wiener process ($D = 1/2$) for a sphere of unit radius, and D is the diffusion coefficient. The distribution of τ_R is well known (e.g. see [116]). Let us present, for generality, the distribution of τ_R for a standard Brownian motion with the diffusion coefficient $D = 1/2$, starting at a distance r from the centre in a sphere of dimension m [116]:

$$H(t, r, R) = \sum_{n=1}^{\infty} \frac{2R^{\nu} J_{\nu}(\mu_n r/R)}{r^{\nu} \mu_n J_{\nu+1}(\mu_n)} \exp\{-\mu_n^2 t/2R^2\}, \tag{8.9}$$

where μ_n, $n = 1, 2, \ldots$ are positive roots of the Bessel function $J_{\nu}(z)$, with $\nu = m/2 - 1$; r is the distance from the starting point x_0 to the centre of the sphere.

In our case of dimension $d = 3$, we have $\nu = 1/2$, and $J_\nu(z) = \sqrt{\frac{2}{\pi z}}\sin(z)$, and $\mu_n = \pi n$, $n = 1, 2\dots$. Besides, $J_{3/2}(z) = \sqrt{\frac{2}{\pi z}}\left(\frac{\sin(z)}{z} - \cos(z)\right)$. Therefore,

$$\frac{2R^\nu J_\nu(\mu_n r/R)}{r^\nu \mu_n J_{\nu+1}(\mu_n)} = -\frac{2R}{\pi r n}\sin(\pi r n/R)(-1)^n.$$

This simplifies the form of the distribution function in dimension 3:

$$H(t,r,R) = -2\sum_{n=1}^{\infty}(-1)^n \frac{R}{\pi r n}\sin(\pi r n/R)\exp\{-\pi^2 n^2 t/2R^2\} \tag{8.10}$$

so by differentiating with respect to t the distribution $1 - H(t,r,R)$, we get the probability density in the form

$$p(t,r,R) = \sum_{n=1}^{\infty}(-1)^{n+1}\frac{\pi n}{rR}\sin(\pi r n/R)\exp\{-\pi^2 n^2 t/2R^2\}. \tag{8.11}$$

In the 2D case,

$$p(t,0,1) = \sum_{k=1}^{\infty}\frac{j_k}{J_1(j_k)}\exp\{-j_k^2 t/2\}. \tag{8.12}$$

Simulation of the exit time τ_1 according to $p(t,0,1)$ was carried out as follows. We have approximated this density by a discrete distribution $p_1, p_2, \dots, p_k$ and used the Walker's algorithm (we have presented an efficient implementation of this method in [94]), which uses only one sample of $rand$ to generate one sample of τ_1.

To simulate the random exit point for a Brownian motion starting in a point at a distance r from the centre, we use the following very simple algorithm.

Let us denote by $\mathbf{x}$ the coordinates of the starting point. First, we simulate a line going through $\mathbf{x}$, having a random isotropic direction. This line intersects the sphere in two points, say, A and B. Let $a = |A - \mathbf{x}|$ and $b = |B - \mathbf{x}|$ be the distances from these points to the starting point $\mathbf{x}$. Then, we choose the point A with probability $b/(a+b)$ and the point B with probability $a/(a+b)$.

This elegant algorithm follows from the next simple arguments. An orthogonal projection of a 3D Wiener process starting from a point on an arbitrary line is also a Wiener process living on this line. It remains to recall that the splitting probabilities of a 1D Wiener process starting from a point x in a segment $[A, B]$ are $b/(a+b)$ and $a/(a+b)$ where b is the distance from x to the left segment's bound A and a is the distance from x to the right bound B (e.g. see [90]). Note that a direct, quite technical proof of this method was given in [45]. This algorithm is very useful when simulating the random walk where the Markov chain points do not coincide with the centres of spheres. See, for instance [31] where the authors have derived a $\log\log\varepsilon$ law for the mean number of steps of this process.

8.3.2 Radiative and nonradiative recombination in the absence of diffusion

Let us now turn to the algorithm description when the radiative and nonradiative recombinations of electrons and holes happen without diffusion. The initial configuration is prepared by sampling, independently, the coordinates of a set of electrons, holes and recombination centres uniformly in the box. So the coordinates are real numbers, and we do not introduce any mesh in space.

At the first step, we have to choose one of the possible events: (*i*) radiative recombination, i.e., direct annihilation of an electron–hole pair, (*ii*) an electron is captured by one of the empty recombination centres N^+, (*iii*) a hole is captured by one of the $N - N^+$ filled recombination centres (if there are any). The latter processes are described by the probabilities $b_n \exp(-|\mathbf{x}|/a_n)$ or $b_p \exp(-|\mathbf{x}|/a_p)$. To unify the notation, we write $B_{ij} = B_0 \exp(-|\mathbf{r}_{ij}|/a)$, where $|\mathbf{r}_{ij}|$ is the distance between the reacting particles. Thus, the simulation algorithm in the absence of diffusion can be described as follows.

(1) Put $t = 0$ and sample $n = n_0$ electrons and $p = n_0$ holes at random, independently and uniformly distributed in the box. Independently, N recombination centres are also sampled uniformly and mutually independently distributed in the box. Without loss of generality, assume that $N = N^+$.

(2) Sample one of the three possible events: (*i*) direct (radiative) electron–hole annihilation, (*ii*) electron capture by an empty centre, (*iii*) hole capture by a filled centre. To do this, first calculate the majorant frequencies for the three events:

$$\lambda_1 = npB_0 \exp\{-\frac{r_{np,min}}{a}\}, \tag{8.13}$$

$$\lambda_2 = nN_+ b_n \exp\{-\frac{r_{nN_+,min}}{a_n}\}, \tag{8.14}$$

$$\lambda_3 = p(N - N_+) b_p \exp\{-\frac{r_{p(N-N_+),min}}{a_p}\}, \tag{8.15}$$

where $r_{np,\min}$ is the minimal of all possible distances between n electrons and p holes in the box, $r_{nN^+,\min}$ the same for the distances between n electrons and N^+ centres, and $r_{p(N-N^+),\min}$ for the distances between p holes and $N - N^+$ centres. From these frequencies, calculate the probabilities p_1, p_2, p_3 of the events (*i*), (*ii*) and (*iii*), respectively: $p_1 = \lambda_1/\lambda$, $p_2 = \lambda_2/\lambda$ and $p_3 = 1 - p_1 - p_2$, where $\lambda = \lambda_1 + \lambda_2 + \lambda_3$.

(3) Using the probabilities p_1, p_2, p_3, sample the respective event $k = i, ii, iii$, calculate the time increment as $\Delta t = -\log(rand))/\lambda$, and calculate $t := t + \Delta t$.

(4) For the sampled event $k(= i, ii, iii)$, choose uniformly the relevant interacting pair and check if the interaction takes place. For instance, if $k = i$, i.e., the sampled event happens to be radiative recombination of an electron–hole pair, calculate $\bar{P}_{np} = \exp\{\frac{-r + r_{np,min}}{a}\}$. If $rand < \bar{P}_{np}$, then the event occurs. Hence, recalculate $n := n - 1$ and $p := p - 1$ and go back to step 2. Otherwise, nothing happens, the

probabilities p_1, p_2, p_3 remain the same and so return to step 3. Here r is the distance in the sampled interacting pair. For the case $k = ii$, when an electron is captured by the recombination centre, put $n := n - 1$, $N^+ := N^+ - 1$, but when a hole is captured by the recombination centre, we have $p := p - 1$, $N^+ = N^+ + 1$.

If the concentrations n and p are calculated at some prescribed time instances t_m, $m = 1, \ldots M$, just score the values $n(t_m), p(t_m)$, $m = 1, \ldots M$. To calculate the photon flux $\phi(t)$, say, at a time t from the interval $t \in [t_1, t_2]$, count M, the number of electron–hole annihilations that have occurred during this time interval, and take the approximation $I(t) \approx M/(t_2 - t_1)$.

(5) To carry out the average, run the steps 1–4 independently, say, ν times, with ν being a sufficiently large number, and take the arithmetic mean.

Note that there is no need to recalculate the value $r_{\min}$ after each annihilation or capture in the recombination centre: this should be done only if the pair with $r_{\min}$ reacts.

8.3.3 General case of radiative and nonradiative recombination in the presence of diffusion

In the general case when the electrons not only recombine but also diffuse, the algorithm is constructed analogously, by introducing an additional event, namely, that a diffusion jump of an electron happens. Frequency of this event is defined by the diffusion frequencies λ_i given above in (8.8). Again, analogously to the recombination event frequencies, we take a majorant for the diffusion events as $\lambda_4 = n/\tau_{\min}$. Now, the total frequency in the system is $\lambda = \sum_{k=1}^{4} \lambda_k$. Thus the algorithm in the general case looks almost the same as described above for the recombinations without diffusion. The only difference is now, that if the event "diffusion" is sampled, we choose the relevant electron from distribution (8.8) and simulate the jump to the surface of its separation sphere; then we check if the new position is close to one of the holes or nonradiative recombination centres in the sense that the distance between the electron and the hole (or the nonradiative centre) is less than a prescribed interaction radius ε. In our calculations, we have taken $\varepsilon = 0.1\text{nm}$. If this happens, we remove this electron and the relevant hole, or, in the case the electron was close to a free nonradiative centre, the electron is captured by this centre and is now in the state of being free to accept a hole. If however the new position after the jump does not satisfy this condition, we have to redefine the new separation sphere of this electron, while other spheres are not changed. This ends up one time step, and the new time is increased by $\Delta t = -\log(rand)/\lambda$.

8.4 Simulation results and comparison

We have carried out a set of numerical simulations for the 2D case in our previous studies (e.g. see [51, 93, 100]), where we used different methods, and the KMC method and its implementation in the form of cellular automata algorithms among others. In this section, we present some further results, and in particular, we include 3D examples and discuss some comparisons with other methods. We focus on the comparison between the conventional KMC method and the random WOS–based KMC algorithm described in this chapter. We also present some comparisons with deterministic methods implemented in the form of explicit FD schemes.

Our simulations are adapted to take place on a spatial and temporal scale comparable to the experiments as discussed in our study [100]. Hence, we use nm and ns as the natural units. Specifically, the characteristic tunnelling distance a is given in nm, the recombination rate B_0 in ns^{-1}, the diffusion coefficient in $\mathrm{nm}^2\,\mathrm{ns}^{-1}$, and the concentration and intensity are calculated for times up to 10^{12} ns.

Initial conditions are set by assuming that the initial concentrations $n_0 = p_0$ as well as the number of nonradiative centres N are all randomly, independently and uniformly distributed in a box of size $L^d\,\mathrm{nm}^2$, $d = 2, 3$. For the Monte Carlo simulations, the results detailed below were obtained by averaging over ν runs, which corresponds to ν independent initial configurations with ν varying between 100 and 1000.

Consider first the simplest case, a pure tunnelling (no centres, no diffusion). Fig. 8.1 compares the concentration and the intensity for 2D and 3D dimensions obtained by the KMC method. It is seen that the long-time asymptotics is well reproduced in the numerical results. Having the same initial concentrations $n_0 = 0.01$, and rates of tunnelling annihilation, $a = 4$ and $B_0 = 0.04$, the concentration of electron–hole pairs is decreasing considerably higher in the 3D case.

In our calculations, due to zero diffusion, we were able to reach the times of about 10^{12}ns. The case of nonzero diffusion (and no tunnelling) is illustrated by the results presented in figure 8.2 for a 2D box. These results were obtained by the random WOS–based KMC method described in this chapter and compared with the results obtained by the conventional KMC algorithm. The initial concentrations were taken as $n_0 = p_0 = 10^{-3}$, so in the KMC algorithm, the diffusion trajectories are very long before electrons meet and annihilate with a hole, in contrast to the electron trajectories in the WOS-based KMC method. Both methods agree well with the long-time asymptotics, if the simulation box is large enough. In our simulations, we have used 500×500, and $10^3 \times 10^3$, which was relatively large. We also compared the intensities for the case when nonradiative centres were introduced into the box, see the two lower curves in figure 8.2.

Figure 8.3 shows transients obtained by the KMC method and a FD approximation for the diffusion equation (Laplace equation) with the inner point boundaries at the hole positions. It is seen that the concentrations obtained by the FD approximation

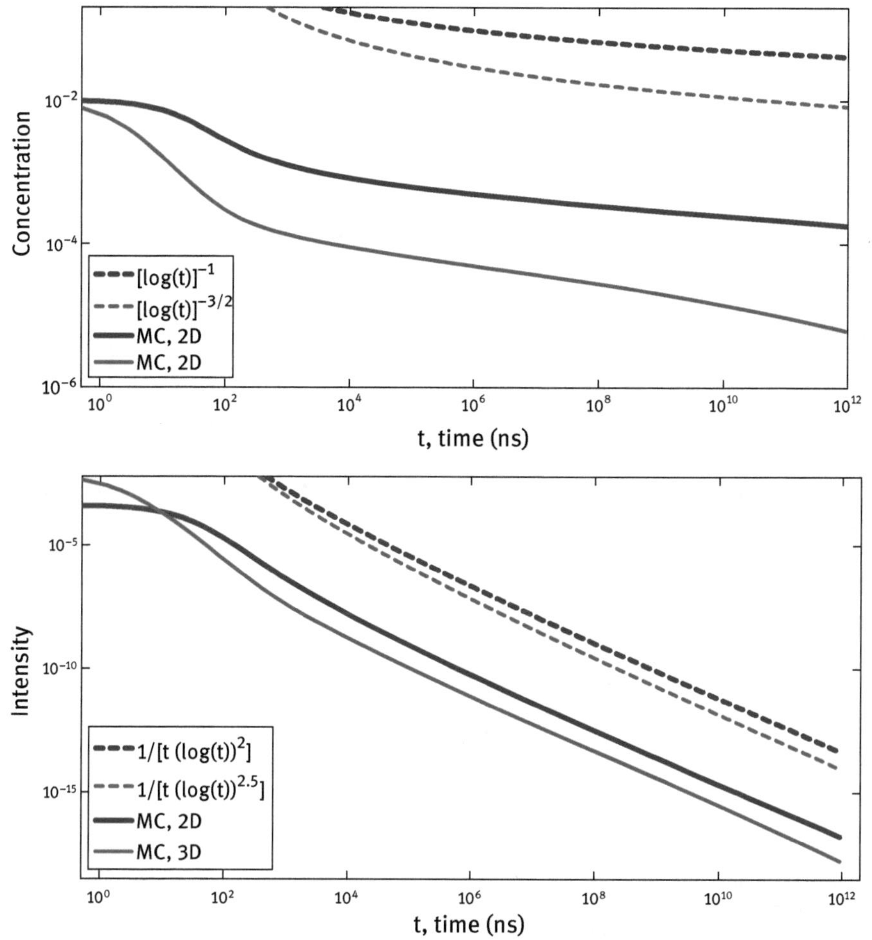

Fig. 8.1. (online colour) Pure tunnelling case: the concentration of electrons (upper panel) and the intensity (lower panel) for $a = 4$, $B_0 = 0.04$.

start to deviate, after the time $t \sim 10^2$, from the concentration obtained by the KMC method. This is caused by the small size of the box we have to take in the FD method.

Finally, the third diffusion model we tried was the Smoluchowski approximation where the annihilation rate due to diffusion was taken in the form $4\pi D\beta$. In Figure 8.3, lower panel, we compare the results obtained by the KMC algorithm for $D = 1$ and $D = 6$ with the Smoluchowski approximation for $\beta = 0.3$. This value was chosen to make as close as possible the intensity functions obtained by these two different methods. It is seen that the results become close to each other only after the time $t \sim 10^3$, i.e., in the large asymptotic range.

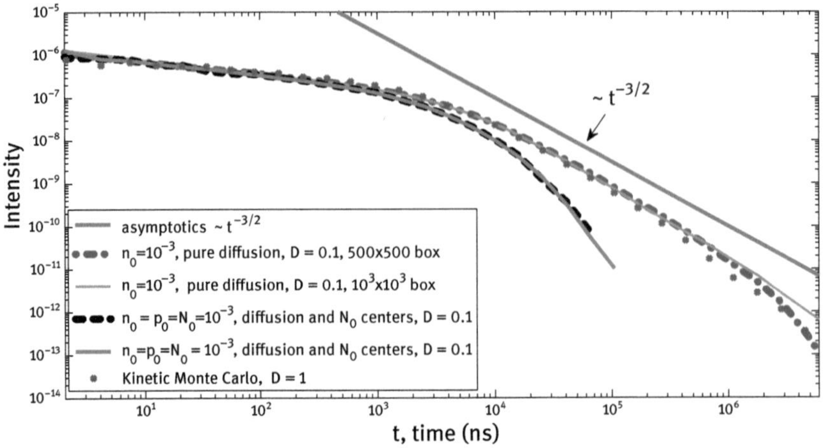

Fig. 8.2. (online colour) Comparison of the intensities obtained by KMC and the combined random WOS and KMC algorithm.

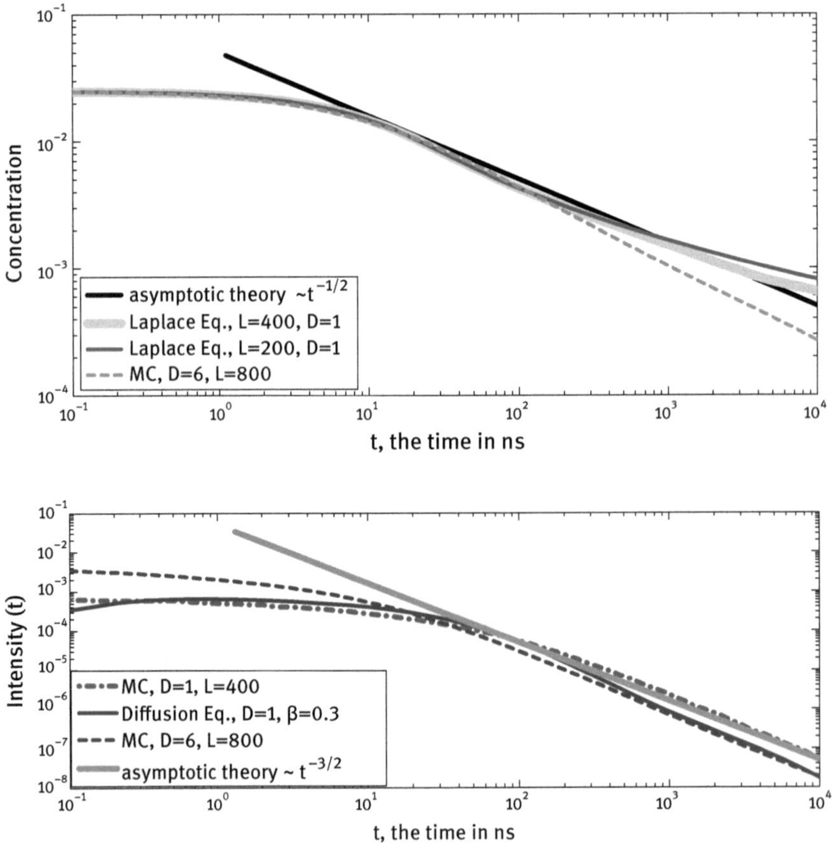

Fig. 8.3. (online colour) Comparison of the concentration (upper panel) and intensity (lower panel), obtained by the KMC algorithm with a finite-difference solutions of the diffusion equations and theoretically predicted long-time asymptotics.

8.5 Summary and conclusion

An extremely fast stochastic algorithm for the simulation of the electron–hole recombination and diffusion dynamics in the presence of nonradiative recombination centres in disordered media is suggested. The method is based on the random walk-on-spheres method combined with the KMC algorithm. To validate the method, we have performed simulations both in 2D and 3D cases and compared the results with the theoretically known long-time asymptotics. The pure diffusion case was analyzed to check other existing models, in particular, the Smoluchowski diffusion approximation.

9 Monte Carlo methods for computing macromolecules properties and solving related problems

Many problems related to calculation of integral physical properties of macromolecules can be reduced to solving BVPs for elliptic equations. Calculation of the diffusion-limited reaction rate, the electrostatic potential and field, the internal energy are all problems that lead to the solution of a diffusion equation with some conditions on the boundary and at the infinity. The intrinsic analogy between diffusion and electrostatics makes it possible to apply the same computational techniques to solve problems coming from these different fields. It is worth noting that the solution domain in this class of problems is usually infinite. So, it is natural that since the days of Maxwell, the boundary-element method has been used as an effective tool for solving such problems. In particular, one can calculate the capacitance and the diffusion-limited reaction rate as a surface integral [65]. The boundary-element method, as well as FD and finite-element methods, is still commonly used for solving electrostatics and diffusion problems arising in biophysics. A review of these and other techniques used in the computation of molecular electrostatic properties is given in [14].

Another possible way of computationally treating these problems comes from the probabilistic representation of solutions to elliptic and parabolic PDEs as functionals of diffusion process trajectories [28, 29, 43]. The direct computational simulation of physical diffusion in this case coincides with the approximation of the Brownian motion as the solution to a stochastic differential equation via a first-order Euler scheme [52, 71]. This approach was applied [22] to simulation studies of diffusion-limited reactions. Though computationally far from being optimal, it allows one to include different physical phenomena (hydrodynamic, electrostatic, etc.) into the computational scheme, and, what is essential, it is efficient enough to be competitive with deterministic methods. Later, this algorithm was modified [124] by incorporating different boundary conditions [105, 115].

It is worth noting that probabilistic representations strongly depend on the type of these conditions. In particular, taking into account reflective surfaces incorporates the local time spent by the diffusion process on the boundary [28]. Up to now, there has been no effective way to construct a computational algorithm based on this representation. So, the only conceivable method to implement the reflective conditions into the Brownian dynamics computations is to simulate them directly.

From here, it follows that methods based on a reformulation of a problem in the form of an integral equation and application of the Monte Carlo technique for constructing estimates for their solution [21, 23, 24, 98, 104] could be the most appropriate and efficient in this case. In particular, the walk-on-boundary and the WOS [76] methods are based on such an approach.

9.1 Diffusion-limited reaction rate and other integral parameters

9.1.1 Formulation of the problem

Let $G \subset \mathbb{R}^3$ be a compact set representing a large molecule. Suppose that its boundary ∂G is a simple connected piecewise Lyapunov surface [34]. The reaction rate of G with small, point-like particles, which move only due to diffusion, is determined by the flux of these particles onto the surface of the body of this molecule:

$$K = -\int\limits_{\partial G} n \cdot J \, ds.$$

Here, n is the normal vector on ∂G pointing out of G. Due to Fick's law, with no external forces present, the flux is given by $J = -D\nabla\rho$, where ρ is the density of particles and D is their diffusion constant [91].

In the stationary case, the mass conservation law, $\operatorname{div} J = 0$, gives $\Delta\rho = 0$, i.e., the density, ρ, satisfies the Laplace equation. Moreover, since it is natural to assume that the concentration of particles far enough away from a large molecule is not affected by its presence and may be considered to be constant, ρ may be set equal to 1 at infinity.

The Laplace equation requires that ρ satisfies a boundary condition on ∂G. Suppose that the reaction rate of particles with the large molecule is proportional to the probability of their encounter (i. e. the particle density near the surface of the molecule). This means $-n \cdot J = \kappa_S \rho$ on ∂G, where $\kappa_S \geq 0$ is the second-order rate constant of a particle absorbed onto the surface: this reaction rate may be anisotropic [91].

In the limit as $\kappa_S \to +\infty$, all the particles absorbed on the surface react immediately. However, the concentration gradient remains finite, so that the density, ρ, must equal zero, and we thus arrive at the Dirichlet boundary condition. However, if $\kappa_S = 0$ at some point, then the flow equals zero at that point. From this, it follows that all the particles must be reflected (the Neumann BVP). In an intermediate general case, there are non-zero probabilities of both absorption (reaction) and reflection.

Therefore, the problem of calculating K can be reduced to an external BVP for the function $u = 1 - \rho$ as follows:

$$\begin{aligned}
&\Delta u(x) = 0, \ x \in G_1 = \mathbb{R}^3 \setminus \overline{G}, \\
&\kappa_S(y)u(y) - D\frac{\partial u}{\partial n}(y) = \kappa_S(y), \ y \in \partial G, \\
&\lim_{|x| \to \infty} u(x) = 0.
\end{aligned} \tag{9.1}$$

To calculate K, we have to reformulate its analytical expression, since we have no exact information on the normal flux $-n \cdot J = D\frac{\partial \rho}{\partial n} = -D\frac{\partial u}{\partial n}$ at the boundary.
Without loss of generality, assume that the origin is in G, and consider $B(0,r)$, a ball centred at the origin of radius r that contains G, with $S(0,r)$ being its surface. By the Green's identity for the domain $B(0,r)\setminus\overline{G}$, we have

$$K = -D \int\limits_{\partial G} \frac{\partial u}{\partial n} d\sigma = D \int\limits_{S(0,r)} \frac{\partial u}{\partial n} d\sigma$$
$$= -\int d\Omega \, Dr^2 \frac{\partial u}{\partial r}(r,\Omega) = -4\pi Dr^2 \frac{\partial \overline{u}}{\partial r},$$

where

$$\overline{u}(r) = \int d\Omega \, \frac{1}{4\pi} u(r,\Omega) \tag{9.2}$$

is the solid angle average of u. Next, we divide both sides of this equality by r^2 and integrate over r from a fixed R to infinity. Finally, we obtain that

$$K = 4\pi RD \, \overline{u}(R). \tag{9.3}$$

It is essential to note that this formula is valid for arbitrary boundary conditions. In fact, this expression was extensively used in calculations [62, 78, 124].

Despite the simplicity of equality (9.3), the reaction rate, K, can be written down analytically only when the molecule is a sphere. Even for the simplest Solc-Stockmayer model [115], when only a circular patch on the surface of a spherical molecule is reactive while the rest of the boundary is inert, some approximations are needed in order to obtain an analytic solution [105].

Different computational methods can be used to solve BVP (9.1) and to calculate the reaction rate in accordance with (9.3). It has been found [78] that the most efficient algorithm is a stochastic one based on the isotropic random choice of a point, x_0, on the sphere, $S(0,R)$. This corresponds to the randomized computation of integral (9.2), i.e., $\overline{u}(R) = \mathbb{E}u(x_0)$.

Next, we use a Monte Carlo estimate $\xi(x_0)$ instead of the exact value, $u(x_0)$. If ξ is independent of x_0, then (for an unbiased ξ) we obtain the following estimate for K:

$$4\pi RD \; \mathbb{E}\,\xi(x_0) = K. \tag{9.4}$$

To construct an estimate $\xi(x_0)$, we can use either walk-on-boundary or the WOS algorithm [21] and its modification for the case of exterior BVPs [24]. The WOS algorithm provides the tool for efficient simulation of exit points of the diffusion process to the domain's boundary. Averaging of known values at these points provides an estimate for the solution. This algorithm works well for the Dirichlet boundary conditions. Common way of treating flux conditions is to simulate reflection from the boundary in accordance with the FD approximation of normal derivative [35, 57, 64, 69]. Such approach introduces a bias into the estimate.

Application of the walk-on-boundary algorithms makes it possible to solve not only Dirichlet but also Neumann and third BVPs. The same approach can be used also for solving problems with continuity boundary conditions [49, 50]. However, this class of algorithms has its limitations.

Next, we consider the WOS-based approach [64]. It has been shown [106] that the most convenient and efficient way to realize WOS (from the point of view of the computational complexity) is to represent the domain as a union of overlapping subdomains each of which is geometrically simple. Such a representation makes it possible to construct the exit point from each subdomain separately and then to either score the boundary value when exit from the domain occurs or to continue a diffusion path inside another subdomain (see Fig. 9.2 as an example).

It is clear that the best possible case is when we know the Green's functions for these subdomains and, therefore, can jump from the interior point directly to a particular subdomain's boundary. In any case, if the mean number of steps for the Markov chain of a diffusion path in every subdomain is bounded, then by a simple extension of the Schwartz lemma, the total number of steps is also bounded and has the same order [106]. For clarity, we expound these results in the next section 9.2.

As an example, let G be a ball $B(0, R_0)$ with an open spherical cavity. This means that the exterior of G can be represented as a union of two domains: the first one is the exterior of a large sphere $\mathbb{R}^3 \setminus B(0, R_0)$ and the second one is the interior of the spherical cavity $B(x_c, R_c)$. The cavity will be open if we assume that $R_0 - R_c < |x_c| < R_0 + R_c$ (Fig. 9.1).

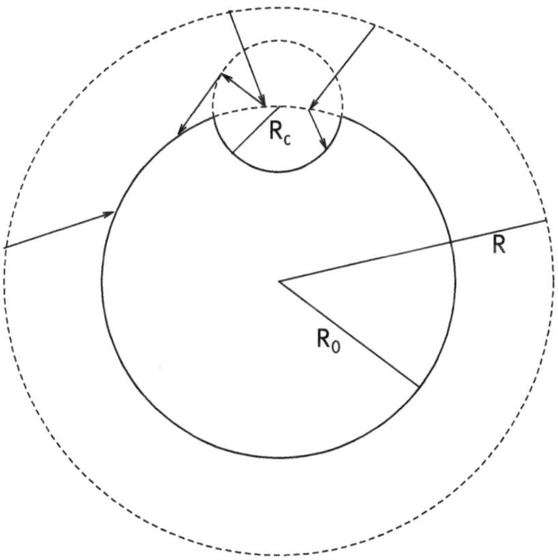

Fig. 9.1. Example of a model computational domain in the reaction rate calculation.

Let $\{x_i, i = 0, 1, \ldots\}$ be a Markov chain of points used for estimating the reaction rate, and let N be its random length. In this case, the chain is constructed as follows. The first point is simulated isotropically on a sphere $S(0, R)$ that contains G: $x_0 = R\omega$, where ω is a unit isotropic vector (R can be taken to be equal to R_0). For $R > R_0$, we apply the algorithm used in the case of the spherical molecule and based on the randomization of the Poisson integral (for $|x| > R_0$)

$$u(x) = \frac{1}{4\pi R_0} \int\limits_{S(0,R_0)} \frac{|x|^2 - R_0^2}{|x-y|^3} u(y) d\sigma(y). \tag{9.5}$$

So, with probability $1 - R/R_0$, the particle moves to infinity ($x_1 = \infty$), $N = 1$ and $\xi(x_0) = 0$. With the complementary probability, R/R_0, the particle is placed on the sphere $S(0, R_0)$. In this case, there are two possibilities. The first is that $|x_1 - x_c| > R_c$. This means that x_1 lies on the boundary of the molecule ∂G. Here, we have to utilize the boundary conditions.

Suppose now that $|x_1 - x_c| < R_c$. In this case, we can use the Poisson kernel for the interior of a ball $B(x_c, R_c)$, which differs from (9.5) only in its sign. So, the next point, x_2, is chosen randomly on $S(x_c, R_c)$ in accordance with the Poisson's density $\frac{1}{4\pi R_c} \frac{R_c^2 - |x_1|^2}{|x_1 - y|^3}$. Part of $S(x_c, R_c)$ coincides with the boundary of the molecule, where we apply the boundary condition. The rest of the surface of sphere $S(x_c, R_c)$ lies outside of $B(0, R_0)$, where we can once more apply formula (9.5) to simulate the next point, x_2, on the surface of the large sphere.

Finally, the constructed Markov chain stops either at the absorbing boundary, where we set $\xi(x_0) = 1$, or at the infinity, where the estimate assumes a value of zero. Clearly, for a fixed x_0, the estimate $\xi(x_0)$ follows the binomial distribution.

Apparently, the same reasoning makes it possible to use this efficient algorithm to construct a Monte Carlo estimate in the case when the molecule, G, can be represented as a sphere with several intersecting spherical cavities.

Note that the described algorithm can also be applied to problems with stochastic gating of a cavity [123].

9.1.2 Capacitance calculations

It is a common knowledge that there exists a deep analogy between mathematical models used in diffusion, heat conduction and electrostatics. This allows using the same approach to constructing computational algorithms for solving problems coming from different areas of physics. In particular, the same random walk simulations had been used for calculating hydrodynamic friction, diffusion limited reaction rate and electrical capacitance (see, e.g. [22, 30, 124]).

Here, we consider the problem of computing the electrostatic capacitance of an isolated body, G. It can be determined by calculation of the surface integral

$$C = \int\limits_{\partial G} -\frac{1}{4\pi}\frac{\partial u}{\partial n}\, d\sigma(y),$$

where u is the solution of the exterior Dirichlet problem for the Laplace equation with the boundary value equal to 1. Using the same reasoning as in Section 9.1.1, we reduce the problem to computation of the following estimate:

$$R\ \mathbb{E}\,\xi(x_0) = C, \tag{9.6}$$

where $\mathbb{E}(\xi(x_0)|x_0) = u(x_0)$ for a random point x_0 uniformly distributed on the spherical surface, $S(0, R)$, of a ball, which completely contains G.

As an example, we consider the classical test problem in electrostatics: computation of capacitance for the unit cube. In Section 6.1, we showed how this problem can be solved by the walk-on-boundary algorithm. Here, we utilize the specifics of the problem's geometry and apply the walk-in-subdomains technique (Section 9.2). In this particular case, the computational domain $G_e = \mathbb{R}^3 \setminus \overline{G}$ can be represented as the unity of seven subdomains: six half-spaces

$$G_{2i-1} = \{x : x_{(i)} < -1/2\}, \quad G_{2i} = \{x : x_{(i)} > 1/2\}, \quad i = 1, 2, 3 \tag{9.7}$$

and the exterior of the starting sphere $S(0, R)$. Here, R must be greater or equal to $\sqrt{3}/2$.

The algorithm of simulating a random walk and its exit points (to the boundary, ∂G, or to infinity) works as follows. It begins with the generation of an isotropically distributed random vector x_0 of length R. We take this point as the starting point of a Markov chain $x^{(0)} = x_0$.

Suppose this chain was simulated up to the point $x^{(i)}$. With probability 1, this point falls into one of the half-spaces (9.7). Let, for definiteness, $x^{(i)} \in G_6$. For a half-space, the Green's function of the Dirichlet problem, $\Phi(x, x^{(i)})$, is well known, and the exit point for the Brownian motion in this domain is distributed in accordance with the density

$$\frac{\partial \Phi}{\partial n(x)} = \frac{1}{2\pi}\frac{x_{(3)} - x^{(i)}_{(3)}}{|x - x^{(i)}|^3} = \frac{1}{2\pi}\frac{z}{(z^2 + r^2)^{3/2}}.$$

Here, $n = (0, 0, -1)$ and we denote by $z = |x_{(3)} - x^{(i)}_{(3)}|$ the distance from $x^{(i)}$ to the plane $\{x_{(3)} = 1/2\}$, and $r = \sqrt{(x_{(1)} - x^{(i)}_{(1)})^2 + (x_{(2)} - x^{(i)}_{(2)})^2}$.

Then, we turn to the polar coordinates and simulate the next point in the Markov chain, $x^{(i+1)} = (x^{(i)}_{(1)} + r\cos\varphi, x^{(i)}_{(2)} + r\sin\varphi, 1/2)$ using the exact formulas $\varphi = 2\pi\alpha_0$,

$r = z\sqrt{\alpha_1^{-2} - 1}$. Here, α are independent uniformly distributed in $[0, 1]$ standard random variables.

At every step of simulation, there is a non-zero probability that $x^{(i)}$ falls into the seventh subdomain, i.e., $|x^{(i)}| > R$. For the exterior of a sphere, the Green's function is exactly known, and we can use the same Poisson formula (9.5) as in the reaction-rate computation. Integration of the Poisson kernel gives $1 - R/|x^{(i)}|$. This means that with probability $R/|x^{(i)}|$, the estimate is set to zero, while with the complement probability, the next point $|x^{(i+1)}|$ is simulated on $S(0, R)$. Setting estimate to zero may be interpreted as the jump of the Markov chain to infinity.

Note that $\mathbb{E}(\xi(x_0)|x_0)$ is equal to probability of a Markov chain initiated at x_0 to terminate on the surface of G. This means the second moment of $\mathbb{E}(\xi(x_0)|x_0)$ equals its expectation, and the variance of $\xi(x_0)$ is bounded.

9.2 Walk in subdomains and efficient simulation of the Brownian motion exit points

Consider the problem of efficient simulation of exit points from a domain G for the Brownian random process.

For simple domains, it is possible to construct an exact representation of the Green's function for the Dirichlet problem, Φ, and simulate exit points on the domain boundary $\Gamma = \partial G$ in accordance with the normal derivative of this function, $\frac{\partial \Phi}{\partial n}$. Note, however, that even if Φ is constructed, the simulation algorithm could be prohibitively complicated and time-consuming. Commonly, the most reasonable solution in such situation is to apply the WOS algorithm (or analogous methods based on simulation of exit points from some other simple domains completely contained in $\overline{G}$), which is stopped either exactly on Γ or in the ε-strip near the boundary.

On the other hand, every step of constructing a random walk inside a domain requires the exact value of the distance from a current point to the boundary, Γ. Calculation of this value is the classical problem of computational geometry, and various approximate nearest neighbour algorithms can be efficiently applied for solving this problem in the case of a complicated surface [2, 3, 15].

One of the possible ways to overcome this algorithmic complexity is based on the representation of the whole domain in the form of a unity of intersecting subdomains [106]. We suppose that in every subdomain, simulation of a Monte Carlo estimate is simple and efficient. In a particular case, exact values of the Green's functions could be used [73].

Consider the Dirichlet BVP for the Laplace equation (3.5) in a domain

$$G = \bigcup_{j=1}^{M} G_j. \tag{9.8}$$

Suppose that
a) in every G_j, there exists the unique solution u_j of the Laplace equation with the Dirichlet boundary condition, g_j;
b) for every u_j, a Monte Carlo estimate can be constructed

$$\xi[u_j](x) = g_j(t_\star) \cdot Q(x; t_\star; t_1, \dots, t_N), \tag{9.9}$$

where $x \in G_j$, $t_\star \in \partial G_j$ and $t_i \in \overline{G_j}, i \geq 1$ is a Markov chain of length N. The weight Q does not depend on g_j.

Suppose that every subdomain $G_j, j = 1, \dots, M$ intersects with others: $G_j \bigcap (\bigcup_{k \neq j} G_k) \neq \emptyset$. Next, its boundary can be represented as $\partial G_j = \Gamma_{j,1} \bigcap \Gamma_{j,2}$, where $\Gamma_{j,1}$ is a part of Γ, and $\Gamma_{j,2}$ lies inside G. Thus, $\partial G = \bigcup_j \Gamma_{j,1}$, and $\Gamma_{j,2} = \bigcup_{k \neq j} \partial G_{j,k}$, where $\partial G_{j,k} = \partial G_j \bigcap G_k$, i.e., the part of the subdomain G_j boundary, which lies inside G_k.

Then, the estimate for the solution of (3.5) (9.8) can be constructed as follows:

$$\zeta[u](x) = g(t_\star^K) \left(\prod_{k=0}^{K} Q(t_\star^{k-1}; t_\star^k; t_1^k, \dots, t_{N_k}^k) \right), \tag{9.10}$$

where we set $t_\star^{-1} \equiv x$. Here, $x \in G_{(0)}$ and $t_1^0, \dots, t_{N_0}^0$ is a random walk in $\overline{G_{(0)}}$ such that $t_\star^0 \in \partial G_{(0)}$. Then, either $t_\star^0 \in \Gamma_{(0),1}$ or $t_\star^0 \in \Gamma_{(0),2}$. In the first case, the Markov chain simulation is terminated, while in the second case, $t_\star^0$ lies inside some other subdomain, which is denoted as $G_{(1)}$. Finally, the process as a whole is terminated at some step, K.

Further, we give the properties of the constructed estimate (see [106]). To prove them, we used an obvious extension of the Schwartz lemma (see, e.g. [12, 111])

Lemma 9.1. *Let G be a union of two intersecting subdomains, G_1 and G_2. Suppose boundaries of these subdomains consist of a finite number of surfaces with continuously changing normal vector, and these boundaries, ∂G_1 and ∂G_2, intersect at angles separated from zero. Let F satisfy the Laplace equation in G_1. Suppose it is continuous in $\overline{G_1}$, equal zero on $\Gamma_{1,1}$ (i.e., on the boundary of G) and equal one on $\Gamma_{1,2}$ (i.e., inside G_2).*

Then, there exists a constant $0 < q < 1$, which depends on G_1 and G_2 only, such that $F \leq q$ on $\Gamma_{2,2}$ (i.e., inside G_1).

Theorem 9.1. *If $\xi[u_j](x)$ defined by (9.9) is an unbiased estimate for a Dirichlet problem in every subdomain G_j, then $\zeta[u](x)$ constructed in accordance with (9.10) is the unbiased estimate for the Dirichlet problem in the whole domain, $G = \bigcup_{j=1}^M G_j$.*

Theorem 9.2. *Suppose estimates $\xi[u_j](x)$ defined by (9.9) are biased:*

$$\mathbb{E}\xi[u_j](x) = u_j(x) + \delta_j(x), \tag{9.11}$$

where $|\delta_j(x)| \leq \varepsilon B_j$ and $B_j = \sup_{x \in G_j} |u_j(x)|$.

Then, the estimate $\zeta(x)$ in the whole domain G is also biased

$$|\mathbb{E}\,\zeta(x) - u(x)| \le \frac{\varepsilon}{1-(\varepsilon+q)}B,$$

where $q = \max_{i,k} q_{i,k}$, $B = \sup_{x\in G} |u(x)|$.

Note that for the WOS-based estimate, weights Q are less than 1 (see [23]). Therefore, the bias of $\zeta(x)$ coincides with the bias of $\xi(x)$. Thus, it is less than $A\omega(\varepsilon)$ for some constant A, where ω is the continuity modulus of $u(x)$.

Theorem 9.3. *Variance of $\zeta(x)$ is bounded.*

9.3 Monte Carlo algorithms for boundary-value conditions containing the normal derivative

Consider now the case of a general-type boundary condition (9.1). One of the possible ways of dealing with such condition could be utilizing the FD approximation to the normal derivative.

Let $y \in \partial G$ and $x_{h,y} = y + nh \in G_1$, where h is a small number, and $n = n(y)$ is the normal vector at y. Then, the finite-difference approximation of the boundary condition (9.1) can be written as

$$u(y) = p_h u(x_{h,y}) + (1 - p_h) + O(h^2), \tag{9.12}$$

where $p_h(y) = \left(1 + \frac{\kappa_s(y)h}{D}\right)^{-1}$. To take this condition into account via Monte Carlo, we randomize it. When a diffusing particle hits the boundary at a point x_i, we calculate $p_h(x_i)$. With probability $1 - p_h$, the particle is absorbed by the molecule and we set $u(x_i) = 1$. With the complementary probability, p_h, we reflect the particle and set its position to $x_{i+1} = x_{h,y}$. Every time a point of the Markov chain $\{x_i, i = 0, 1, \dots, N\}$ hits the boundary, equation (9.12) is used, which introduces an $O(h^2)$ bias. The resulting bias of the estimate for the reaction rate can be evaluated analytically in the simplest cases only.[1] Generally, the number of returns to the surface of a molecule is $O(1/h)$, which means that the resulting bias is $O(h)$.

9.3.1 WOS algorithm for mixed boundary-value conditions

As an alternative to the FD approximation to flux conditions, we can use an exact integral formula for a function value at a point on the boundary [108, 109].

1 For example, let $\kappa_s = \text{const}$, then the mean number of steps until the particle is absorbed is equal to $\frac{(2D+3\kappa_s h)(R_0+h)}{h(\kappa_s(R_0+h)+D)}$. Hence, the resulting bias is of order h. However, if $\kappa_s h \gg D$, then it is $O(h^2)$.

Consider first the linearized Poisson-Boltzmann (PB) (Helmholtz) equation

$$\Delta u - \kappa^2 u = 0, \ \kappa = \text{const} \geq 0 \tag{9.13}$$

in a bounded domain $G \subset \mathbb{R}^m$, and a mixed BVP for it:

$$\alpha(y)\frac{\partial u}{\partial n}(y) + \beta(y)u(y) = g(y), \ y \in \Gamma = \partial G. \tag{9.14}$$

Here, $\alpha = 1$, $\beta = 0$ on Γ_0, and vice versa: $\alpha = 0$, $\beta = 1$ on $\Gamma_1 = \Gamma \setminus \Gamma_0$. We suppose the boundary is piecewise smooth and that the parameters of the problem guarantee the unique solution exists [72]. To simplify the inference and formulas involved, in the rest of the chapter, we consider only the 3D Euclidean space. It is clear, however, that the integral formula derived stays valid for an arbitrary $m \geq 2$.

To find the solution at a fixed point, x_0, we use the WOS algorithm. Let $x \in G$. We define $d(x)$ as the distance from this point to the boundary Γ. Next, we consider the ball $B(x, d(x))$ and write down the integral Green's formula for the solution, u, and the Green's function for this ball, $\Phi_{\kappa,d}$:

$$u(x) = \int\limits_{S(x,d(x))} \frac{\partial \Phi_{\kappa,d}}{\partial n} \, u \, ds. \tag{9.15}$$

Here, $\Phi_{\kappa,d}(x, y) = -\frac{1}{4\pi}\frac{\sinh(\kappa(d-|y-x|))}{|y-x|\sinh(\kappa d)}$, and $S(x, d(x))$ denotes the sphere of radius $d(x)$ centred at the point x.

Randomization of (9.15) leads to the estimate based on simulation of the WOS Markov chain. The chain is defined by the recursive relation: $x_{i+1} = x_i + d(x_i)\ \omega_i$, where $\{\omega_0, \omega_1, \ldots\}$ is a sequence of independent isotropic unit vectors.

From (9.15), we have $u(x_i) = \mathbb{E}(q(\kappa, d(x_i))u(x_{i+1})|x_i)$, and we treat the factor,

$$q(\kappa, d(x_i)) = \frac{\kappa d}{\sinh(\kappa d)}, \tag{9.16}$$

as the survival probability. For $\kappa = 0$ (or if we use the weight estimate [21]), WOS with probability 1 converges to the boundary. Let x_k be the first point of the Markov chain that hits Γ_ε, the ε-strip near the boundary, and denote by $x_k^\star \in \Gamma$ the point nearest to x_k. Then,

$$u(x_0) = \mathbb{E}(u(x_k)\chi),$$

where $\chi = 0$, if the trajectory was terminated inside the domain, $\chi = 1$, if the Markov chain reached Γ_ε and $\chi = \prod_{i=0}^{k} q(\kappa, d(x_i))$, if we use the weight estimate.

For $x_k^\star \in \Gamma_1$, we have

$$u(x_k) = g(x_k^\star) + \phi_1(x_k, x_k^\star), \tag{9.17}$$

where $\phi_1(x_k, x_k^\star)$ is $O(\varepsilon)$ for elliptic $x_k^\star$.

For $x_k^* \in \Gamma_0$, the boundary conditions give

$$u(x_k) = u(x_k^*) - g(x_k^*)d(x_k) + \phi_0(x_k, x_k^*), \tag{9.18}$$

where $\phi_0(x_k, x_k^*) = O(\varepsilon^2)$ as $\varepsilon \to 0$.

To be sure that the nearest point on the boundary, x_k^*, is elliptic, we use the walk-in-subdomains technique [106].

Note that for $x_k^* \in \Gamma_0$, the value of $u(x_k^*)$ is not known. To estimate it, we will use the integral formula derived in the next section.

9.3.2 Mean-value relation at a point on the boundary

For an elliptic point on the boundary, $x \in \Gamma_0$, consider the ball $B(x,a)$ and the Green's function $\Phi_{\kappa,a}$ for this ball taken at its centre. For every $y \neq x$, we have $\Delta_y \Phi_{\kappa,a} - \kappa^2 \Phi_{\kappa,a} = 0$.

Denote by $B_i(x,a) = B(x,a) \bigcap G$ the interior part of the ball that lies inside the domain, and let $S_i(x,a)$ be the interior part of its surface. We apply the Green's formula to the pair of functions, u and $\Phi_{\kappa,a}$, in $B_i(x,a) \setminus B(x,\varepsilon)$, excluding a small neighbourhood of the point. Both functions satisfy the PB equation in this domain, $\Phi_{\kappa,a} = 0$ on $S(x,a)$, and we suppose that everywhere on $\Gamma \bigcap B(x,a)$, u satisfies the Neumann boundary conditions. Next, we take into account smoothness of u and that for elliptic points, the surface area of $S_i(x,\varepsilon)$ is $2\pi\varepsilon^2(1 + O(\varepsilon))$ when $\varepsilon \to 0$. As a consequence, we obtain in the limit the following integral formula:

$$u(x) = \int\limits_{\partial B_i(x,a)\setminus\{x\}} 2 \frac{\partial \Phi_{\kappa,a}}{\partial n} u \, ds - \int\limits_{\Gamma \bigcap B(x,a)\setminus\{x\}} 2\, \Phi_{\kappa,a}\, g \, ds. \tag{9.19}$$

Clearly, this formula stays valid when $\kappa = 0$. In this case, the Green's function is $\Phi_{0,a}(x,y) = -\frac{1}{4\pi}\left(\frac{1}{|y-x|} - \frac{1}{a}\right)$.

For convenience, we explicitly isolate singularities in the kernels of integral operators and rewrite (9.19) in the following form:

$$u(x) = \int\limits_{\partial B_i(x,a)\setminus\{x\}} \frac{1}{2\pi} \frac{\cos\varphi_{yx}}{|y-x|^2} Q_{\kappa,a}\, u(y)\, ds(y) + \int\limits_{\Gamma \bigcap B(x,a)\setminus\{x\}} \frac{1}{2\pi|y-x|}\left(1 - \frac{|y-x|}{a}\right) Q^1_{\kappa,a} g(y)\, ds(y). \tag{9.20}$$

Here, $\cos\varphi_{yx}$ is the angle between the external (with respect to $B_i(x,a)$) normal vector at a point y and vector $y-x$. The weight function,

$$Q_{\kappa,a}(|y-x|) = \frac{\sinh(\kappa(a-|y-x|)) + \kappa|y-x|\cosh(\kappa(a-|y-x|))}{\sinh(\kappa a)},$$

is smooth, and $Q_{\kappa,a} = q(\kappa,a) = \frac{\kappa a}{\sinh(\kappa a)}$ on the surface of the auxiliary sphere, $S(x,a)$. Clearly, everywhere in $\overline{B}(x,a)\setminus\{x\}$, this weight function is positive and its value is less than 1. For $\kappa = 0$, it equals 1 identically.

The second weight function, $Q^1_{\kappa,a} = \frac{\sinh(\kappa(a-|y-x|))}{a-|y-x|}\frac{a}{\sinh(\kappa a)}$, is also smooth. It is less than or equal to 1, and $Q^1_{\kappa,a} \geq \frac{\kappa a}{\sinh(\kappa a)}$. Obviously, for $\kappa = 0$, it equals 1 identically.

9.3.3 Construction of the algorithm and its convergence

Suppose first that the part of the domain's boundary with the Neumann conditions, Γ_0, is convex. In this case, the kernel of the integral operator in (9.20) is sub-stochastic, which means that it is non-negative and its integral is less than or equal to 1. Therefore, we can use this kernel as the transition density. The term $\frac{1}{2\pi}\frac{\cos\varphi_{x_{i+1}x_i^*}}{|x_{i+1}-x_i^*|^2}$ corresponds to the isotropic distribution of $x_{i+1} \in \partial B_i(x_i^*,a)$ in a solid angle with its vertex at x_i^*. The weight, $Q_{\kappa,a}(|x_{i+1}-x_i^*|)$, is treated as the survival probability on this transition. (Note that for the plane boundary, with probability 1, x_{i+1} is distributed isotropically on the half-sphere $S_i(x_i^*,a)$, and the survival probability equals $q(\kappa,a)$).

The described construction of a Markov chain corresponds to the so-called direct simulation of an integral equation [23]. This means that the resulting estimate for the solution's value at a point, $x = x_0 \in \overline{G}$, is

$$\xi[u](x) = \sum_{i=0}^{N} \xi[F](x_i). \tag{9.21}$$

Here, N is the random length of the Markov chain, and $\xi[F]$ are estimates for the right-hand side of the integral equation. This function is defined by (9.17), (9.18), (9.20), which gives

$$\begin{aligned}
F(x) &= 0, \text{ when } x \in G\setminus\overline{\Gamma}_\varepsilon;\\
&= \int\limits_{\Gamma\cap B(x^*,a)\setminus\{x^*\}} \frac{1}{2\pi|y-x^*|}\left(1-\frac{|y-x^*|}{a}\right)Q^1_{\kappa,a}\, g(y)\, ds(y)\\
&\quad -g(x^*)\, d(x) + \phi_0(x,x^*), \text{ when } x \in \overline{\Gamma}_\varepsilon \text{ and } x^* \in \Gamma_0;\\
&= g(x^*) + \phi_1(x,x^*), \text{ when } x \in \overline{\Gamma}_\varepsilon \text{ and } x^* \in \Gamma_1.
\end{aligned} \tag{9.22}$$

For Markov chains based on the direct simulation, the finiteness of the mean number of steps, $\mathbb{E}N < \infty$, is equivalent to convergence of the Neumann series for the correspondent integral operator, which kernel coincides with the transition density of this Markov chain. Besides that, the kernel of the integral operator that defines the second moment of the estimate is also equal to this density [23]. This means that for an exactly known free term, F, the estimate (9.21) is unbiased and has finite variance. The same is true if the estimates, $\xi[F](x_i)$, are unbiased and have uniformly in x_i bounded second moments. It is clear that we can easily choose such a density that the estimate for the integral in (9.22) will have the requested properties.

To prove that the mean number of steps is finite, we consider the auxiliary BVP:

$$\Delta p_0 - \kappa^2 p_0 = 0, \quad p_0|_{\Gamma_1} = 0, \quad \frac{\partial p_0}{\partial n}|_{\Gamma_0} = 1. \tag{9.23}$$

For plane or spherical Γ_0, the integral in (9.22) equals $(\cosh(\kappa a) - 1)/(\kappa \sinh(\kappa a))$. More generally, for elliptic points, it equals $\frac{a}{2}\left(1 - \frac{(\kappa a)^2}{24} + O(\kappa a)^4\right)$ as $\kappa a \to 0$. For $a/R \to 0$, we take $\varepsilon = O(a/2R)^3$, where R is the minimal curvature radius at the point $x_i^* \in \Gamma_0$, and get $F(x_i) = \frac{a}{2}\left(1 + O\left(\frac{a}{2R}\right)\right)$.

Consider the estimate (9.21) for p_0. Note that the mean-value relation (9.20) is valid only when $\Gamma \bigcap B(x_i^*, a) \subset \Gamma_0$. From here, it follows that radii of auxiliary spheres, a_i, cannot be greater than the distance from x_i^* to the boundary between Γ_1 and Γ_0 and can be arbitrary small. We take it as the maximal possible and calculate (for $\kappa = 0$) the termination probability for such random walk. For small distances, it can be easily shown that the probability, q, of a reflected diffusion coming to Γ_1 and thus being terminated is separated from zero. Therefore, the mean number of reflections near the line that separates Γ_0 and Γ_1 is bounded by some constant, c^*/q. Next, we fix some value, a^*, and divide the set of all reflections into two classes: in the first one, we put reflections with radii, a_i, which are greater than a^*, and the second class includes the reflections with $a_i < a^*$. Let $N_{0,1}$ and $N_{0,2}$ be, respectively, mean numbers of reflections in the correspondent classes. Then, we have $p_0 > N_{0,1}a^*/2$ and, hence, the overall number of reflections, $N_0 = N_{0,1} + N_{0,2}$, is less than $C_a = 2p_0/a^* + c^*/q$.

From here, it follows that the overall mean number of steps of the described version of the WOS algorithm is of the same order as for the case of the Dirichlet boundary conditions, which means $\mathbb{E}N < \mathrm{const}|\log(\varepsilon)|$.

In the previous reasoning, we supposed the error functions ϕ_0 and ϕ_1 to be known exactly. This presupposition provided us with a possibility of proving that the estimate (9.21) is unbiased. To obtain a functioning estimate, we are forced to get rid of these unknowns. The resulting bias in the estimate is $\phi_1 + N_0\phi_0$, which is $O(\varepsilon)$ when $\varepsilon \to 0$. This means that we proved the following proposition.

Theorem 9.4. *The new version of the WOS algorithm provides the estimate, $\xi[u](x)$, for mixed BVP (9.13), (9.14). The variance of this estimate is bounded, and its bias is $O(\varepsilon)$, where ε is the width of the strip near the boundary of the domain. For the required accuracy of solution, $\delta = \varepsilon$, the computational cost of the algorithm is $O(\log(\delta)\ \delta^{-2})$.*

For $\kappa > 0$ and for pure Neumann BVP, the algorithm has the same properties.

Note that with the FD approximation for the solution's normal derivative, every hit of the ε-strip near the boundary, Γ_0, introduces $O(\varepsilon + h^2)$ bias into the estimate (h is the step value in the normal derivative approximation) [57, 69]. The mean number of reflections in this case is $O(h^{-1})$. It means that if $\varepsilon \sim h^2$, then the resulting bias is $O(h)$, and the overall mean number of steps is $O(\log(h)\ h^{-1})$. Thus, for the required accuracy δ, we have to take $\varepsilon \sim \delta^2$ and $h \sim \delta$. The computational cost of this algorithm is $O(\log(\delta)\ \delta^{-3})$.

So, as we see, the approach we described here makes it possible to substantially improve the efficiency of the WOS algorithm when applied to solving mixed and Neumann BVPs.

The algorithm described above works for a convex Γ_0. If it is not the case, the kernel of the integral operator in (9.20) can be negative, and the direct simulation is not feasible. This also means that changing the kernel of the integral operator to its modulus can lead to non-convergent Neumann series, and, thus, to non-operational Monte Carlo estimates.

To solve this problem, we propose using a simple approximation to the integral relation. Consider the tangent plane at the point, $x_i^* \in \Gamma_0$, and simulate the next point of the Markov chain isotropically on the half-sphere, $S^-(x_i^*, a)$, which lies inside the domain. It can be easily shown that for small a/R, such algorithm introduces an $O(a/2R)^3$ error. Therefore, the resulting bias is $O(a/2R)^2$, and the computational cost of such biased algorithm is $O(\log(\delta)\ \delta^{-5/2})$.

9.4 Continuity BVP

We consider the BVP for a function, $u(x)$, that satisfies different elliptic equations with constant coefficients inside a bounded simple connected domain, $G_i \subset \mathbb{R}^3$, and in its exterior, $G_e = \mathbb{R}^3 \setminus \overline{G_i}$. In Section 3.7, we considered such problem and constructed the walk-on-boundary algorithm for its solution. Here, we describe the alternative Monte Carlo method, which is based on simulation of the WOS Markov chain.

Denote, for convenience, by $u_i(x)$ and $u_e(x)$ the restrictions of function $u(x)$ to G_i and G_e, respectively. Let the first function satisfy the Poisson equation

$$\epsilon_i \Delta u_i = -4\pi\rho, \tag{9.24}$$

and the second one satisfy the linearized PB equation

$$\epsilon_e \Delta u_e - \epsilon_e \kappa^2 u_e = 0, \tag{9.25}$$

where $\epsilon_e \geq \epsilon_i$ and κ could be equal to zero. The continuity conditions on the piecewise smooth boundary, Γ, relate limiting values of the solutions, and their fluxes as well:

$$u_i(y) = u_e(y),\ \epsilon_i \frac{\partial u_i}{\partial n}(y) = \epsilon_e \frac{\partial u_e}{\partial n}(y),\ y \in \Gamma. \tag{9.26}$$

Here, the normal vector, n, is pointed out into G_e, and $u_e(x) \to 0$ as $|x|$ goes to infinity. We assume that the parameters of the problem guarantee the unique solution exists [72]. Problems of this kind arise in various scientific applications such as diffusion, heat conduction, geophysics, transport theory, etc., where the solution of an elliptic equation with piecewise constant coefficients should be found in a non-uniform computational domain. In the molecular biophysics applications [14], such problems arise when the electrostatic properties of macromolecules are needed to be computed. In this case, G_i can be thought of as a molecule in aqueous solution. In the framework of the implicit solvent model, only the geometric structure of this molecule is described explicitly, whereas the surrounding water with ions dissolved is considered a continuous medium.

9.4.1 Monte Carlo method

To solve numerically the problem (9.24), (9.25), (9.26), we propose to use a Monte Carlo method [109, 110]. There are several reasons for such a choice. FD and other deterministic computational methods that are commonly used for solving elliptic BVPs encounter apparent complications when applied to calculating electrostatic properties of molecules in solvent. Most of these difficulties are caused by the complexity of molecular surface. On the other hand, the most efficient and commonly used Monte Carlo methods such as the WOS algorithm [21, 23, 76] and the random walk-on-boundary algorithm [104] can analytically take the geometry into account. The latter can be applied to solving not only Dirichlet, Neumann and third BVPs but also for the problems with continuity boundary conditions in the case when $\kappa = 0$ [49, 50]. This method works well for small molecules but becomes computationally expensive for larger structures, which means that it needs substantial optimization and further algorithmic development. It is well known that the WOS algorithm is designed to work with the Dirichlet boundary conditions. With this method, the common way of treating flux conditions is to simulate reflection from the boundary in accordance with the FD approximation to the normal derivative [35, 57, 64, 69]. Such an approach has a drawback. It introduces a bias into the estimate and substantially elongates simulation of Markov chain trajectories.

In [107], we proposed a new approach to constructing Monte Carlo algorithms for solving elliptic BVPs with flux conditions. This approach is based on the mean-value relation written for the solution value at a point, which lies exactly on the boundary. It provides a possibility to get rid of the bias when using Green's function–based random walk methods and treating algorithmically boundary conditions that involve normal derivatives.

Consider the problem of computing the solution to (9.24), (9.25), (9.26) at a fixed point, x_0. Suppose for definiteness that $x_0 \in G_i$. To estimate $u(x_0)$, we represent it as a sum of the regular part and the volume potential:

$$u(x_0) = u^0(x_0) + g(x_0).$$

Here, $g(x_0) = \int_{G_i} \frac{1}{\epsilon_i} \frac{1}{|x_0-y|} \rho(y)dy$. In the molecular electrostatic problems, charges are considered to be concentrated at a finite number of fixed points. Hence, $g(x_0) = \sum_{m=1}^{M} \frac{1}{\epsilon_i} \frac{1}{|x_0-x_{c,m}|} \rho_m$. Usually, in these problems, x_0 coincides with one of $x_{c,m}$.

The regular part of the solution satisfies the Laplace equation in G_i. Therefore, we have a possibility to use the WOS algorithm to find it. Let x_0 be the starting point and $d(x_i)$ be the distance from the given point, x_i, to the boundary, Γ. Generally, WOS Markov chain is defined by the recursive relation: $x_{i+1} = x_i + d(x_i)\omega_i$, where $\{\omega_0, \omega_1, \dots\}$ is a sequence of independent isotropic unit vectors. With probability 1, this chain converges to the boundary [21]. Let x_k be the first point of the Markov chain that hits Γ_ε, the ε-strip near the boundary. Denote by $x_k^* \in \Gamma$ the point nearest to x_k. Clearly, the sequence $\{u^0(x_i), i = 0, 1, \dots\}$ forms a martingale. Therefore,

$$u^0(x_0) = \mathbb{E}u^0(x_k) = \mathbb{E}(u(x_k) - g(x_k)) = \mathbb{E}(u(x_k^*) - g(x_k^*) + \phi),$$

where $\phi = O(\varepsilon)$, for elliptic boundary points, x_k^*.

The mean number of steps in the WOS Markov chain before it for the first time hits the ε-strip near the boundary is $O(\log \varepsilon)$ [21]. In the molecular electrostatics problems, however, it is natural to use the more efficient way of simulating exit points of Brownian motion on the boundary, Γ. The algorithm is based on the walk-in-subdomains approach [106]. Such construction utilizes the commonly used representation of a molecular structure in the form of a union of intersecting spheres (Fig. 9.2). This version of the WOS algorithm converges geometrically, and there is no bias in the estimate, since the last point of the Markov chain lies exactly on Γ.

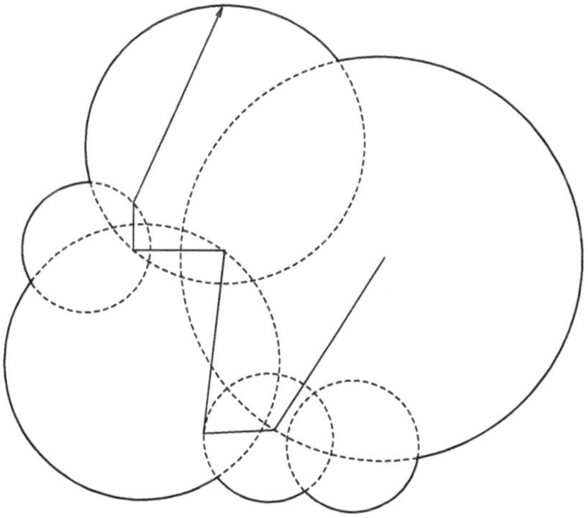

Fig. 9.2. Example of a model geometry and walk in subdomains in the energy calculation.

Note that to compute the estimate for $u^0(x_0)$, we use the unknown boundary values of the solution. With the Monte Carlo calculations, we can use estimates instead of these values. In the next section, we derive the mean-value relation, which is then used in Section 9.4.3 to construct such an estimate.

9.4.2 Integral representation at a boundary point

To elucidate the approach we use, we consider first the exterior Neumann problem for the PB equation (9.25) in G_e:

$$\frac{\partial u_e}{\partial n}(y) = f(y),\ y \in \Gamma. \tag{9.27}$$

Let $x \in \Gamma$ be an elliptic point on the boundary. To construct an integral representation for the solution value at this point, consider the ball, $B(x, a)$, of radius a, and x being its centre.

Let $\Phi_{\kappa,a}(x, y) = -\frac{1}{4\pi}\frac{\sinh(\kappa(a-|x-y|))}{|x-y|\sinh(\kappa a)}$ be the Green's function of the Dirichlet problem for PB equation (9.25) considered in the ball, $B(x, a)$, and taken at the central point of this ball. Denote by $B_e(x, a) = B(x, a)\bigcap G_e$ the external part of the ball, and let $S_e(x, a)$ be the part of the spherical surface that lies in G_e. Next, we exclude from this ball a small vicinity of the point, x. From here, it follows that, for arbitrary $\varepsilon < a$, both functions, u_e and $\Phi_{\kappa,a}$, satisfy the PB equation in $B_e(x, a) \setminus B(x, \varepsilon)$. Therefore, it is possible to use the Green's formula for this pair of functions in this domain. Taking the limit of this formula as $\varepsilon \to 0$, we have

$$\begin{aligned} u_e(x) = &\int_{S_e} 2\frac{\partial \Phi_{\kappa,a}}{\partial n}\ u_e\ ds \\ &- \int_{\Gamma\bigcap B(x,a)\setminus\{x\}} 2\frac{\partial \Phi_{\kappa,a}}{\partial n}\ u_e\ ds \\ &+ \int_{\Gamma\bigcap B(x,a)\setminus\{x\}} 2\Phi_{\kappa,a}\frac{\partial u_e}{\partial n}\ ds. \end{aligned} \tag{9.28}$$

Here, in the second and third integrals, we took into account that, on Γ, the normal vector external with respect to $B_e(x, a) \setminus B(x, \varepsilon)$ has the direction opposite to the normal vector we use in boundary conditions (9.27). The normal derivative of the Green's function can be written down explicitly.

We have $\frac{\partial \Phi_{\kappa,a}}{\partial n}(y) = Q_{\kappa,a}(r)\frac{\partial \Phi_{0,a}}{\partial n}(y)$.

Here, $Q_{\kappa,a}(r) = \frac{\sinh(\kappa(a-r)) + \kappa r\cosh(\kappa(a-r))}{\sinh(\kappa a)} < 1$, $r = |x-y|$,

and $2\frac{\partial \Phi_{0,a}}{\partial n}(y) = \frac{1}{2\pi}\frac{\cos\phi_{yx}}{|x-y|^2}$, where ϕ_{yx} is the angle between $n(y)$ and $y - x$. $\Phi_{0,a}(x, y) = -\frac{1}{4\pi}\left(\frac{1}{|x-y|} - \frac{1}{a}\right)$ is the Green's function for the Laplace equation.

Application of the Green's formula to the pair of functions, u_i and $\Phi_{\kappa,a}$, in $B_i(x,a) \setminus B(x,\varepsilon)$ provides the analogous result. To have a possibility to do this, we have to suppose that there are no charges in this part of the interior domain. From here, it follows that the total potential, u_i, satisfies the Laplace equation. Thus, using the Green's formula and taking the limit as $\varepsilon \to 0$, we obtain

$$\begin{aligned} u_i(x) &= \int_{S_i} 2\frac{\partial \Phi_{\kappa,a}}{\partial n} u_i \, ds \\ &+ \int_{\Gamma \bigcap B(x,a) \setminus \{x\}} 2\frac{\partial \Phi_{\kappa,a}}{\partial n} u_i \, ds \\ &+ \int_{B_i(x,a)} [-2\kappa^2 \Phi_{\kappa,a}] u_i \, dy \\ &- \int_{\Gamma \bigcap B(x,a) \setminus \{x\}} 2\Phi_{\kappa,a} \frac{\partial u_i}{\partial n} \, ds. \end{aligned} \tag{9.29}$$

Note the additional volume integral in this representation.

To make use of continuity boundary conditions (9.26), we multiply (9.28) by ϵ_e and (9.29) by ϵ_i, respectively, and sum up the results. This gives us

$$\begin{aligned} u(x) &= \frac{\epsilon_e}{\epsilon_e + \epsilon_i} \int_{S_e(x,a)} \frac{1}{2\pi a^2} \frac{\kappa a}{\sinh(\kappa a)} u_e \, ds \\ &+ \frac{\epsilon_i}{\epsilon_e + \epsilon_i} \int_{S_i(x,a)} \frac{1}{2\pi a^2} \frac{\kappa a}{\sinh(\kappa a)} u_i \, ds \\ &- \frac{\epsilon_e - \epsilon_i}{\epsilon_e + \epsilon_i} \int_{\Gamma \bigcap B(x,a) \setminus \{x\}} \frac{1}{2\pi} \frac{\cos \phi_{yx}}{|x-y|^2} Q_{\kappa,a} u \, ds \\ &+ \frac{\epsilon_i}{\epsilon_e + \epsilon_i} \int_{B_i(x,a)} [-2\kappa^2 \Phi_{\kappa,a}] u_i \, dy. \end{aligned} \tag{9.30}$$

In the limiting case, $\kappa = 0$, this relation simplifies

$$\begin{aligned} u(x) &= \frac{\epsilon_e}{\epsilon_e + \epsilon_i} \int_{S_e(x,a)} \frac{1}{2\pi a^2} u_e \, ds \\ &+ \frac{\epsilon_i}{\epsilon_e + \epsilon_i} \int_{S_i(x,a)} \frac{1}{2\pi a^2} u_i \, ds \\ &- \frac{\epsilon_e - \epsilon_i}{\epsilon_e + \epsilon_i} \int_{\Gamma \bigcap B(x,a) \setminus \{x\}} \frac{1}{2\pi} \frac{\cos \phi_{yx}}{|x-y|^2} u \, dy. \end{aligned} \tag{9.31}$$

Note that for a plane boundary, $\cos \phi_{yx} = 0$, and the integral over Γ vanishes.

9.4.3 Estimate for the boundary value

To construct an estimate for the solution of the BVP, we need an estimate for the unknown boundary value. To do this, we use the mean-value relation constructed in the previous section. It is not so simple as it may seem at the first sight, since we have to iterate integral operators standing in the right-hand sides of these representations. The kernels of these operators can be alternating, and the convergence of the Neumann series after replacing the kernel by its modulus cannot be ensured. As a consequence, the direct randomization of integral relations (9.28), (9.30), (9.31) cannot be used to constructing a Monte Carlo estimate [23, 24]. However, as we see, the only part of the integral operator, which can be negative, is the integral over the boundary. Thus, if Γ consists of planes, then the direct randomization is really possible.

In the general case, we can use a simple probabilistic approximation. Let $x = x_k^\star$ be the point on the boundary nearest to the last point of the already constructed random walk.

Consider first the exterior Neumann problem. If Γ is concave everywhere in $B(x, a)$, then the kernel of the integral operator in (9.28) is positive, and the next point of the Markov chain, x_{k+1}, can be sampled isotropically in a solid angle with x being its vertex. However, it would be more natural to suppose that the boundary was convex. In this case, we draw the tangent plane to Γ at this point and sample x_{k+1} isotropically on the external hemisphere, $S^+(x_k^\star, a)$. Then, for sufficiently smooth function, u, we have

Lemma 9.2.

$$u(x_k^\star) = \mathbb{E}\left\{u(x_{k+1}) \mid x_k^\star\right\} + \phi_\Gamma,$$

where $\phi_\Gamma = O\left(\frac{a}{2R}\right)^3$ *as* $a/2R \to 0$. *Here, R is the minimal radius of curvature at the point,* $x_k^\star$.

This statement is easily verified by expansion of u into series and direct integration.

The same approach works in the case of continuity boundary conditions (9.26). For $\kappa = 0$, with probability $p_e = \frac{\epsilon_e}{\epsilon_i + \epsilon_e}$, we sample the next point of the Markov chain, x_{k+1}, isotropically on $S^+(x_k^\star, a)$. With the complementary probability, p_i, the next point is chosen on the other hemisphere, $S^-(x_k^\star, a)$.

If $\kappa > 0$, with probability p_e, we sample the next point on $S^+(x_k^\star, a)$ and treat the coefficient $q(\kappa, a) = \frac{\kappa a}{\sinh(\kappa a)}$ as the survival probability. With probability p_i, we sample the direction of the vector, $x_{k+1} - x$, isotropically pointing to $S^-(x_k^\star, a)$. Next, with probability $q(\kappa, a)$ ($Q_{\kappa,a}$, if this vector intersects Γ), the next point is taken on the surface of the sphere, and with the complementary probability, it is sampled inside the ball, $B^-(x_k^\star, a)$. The simulation density of $r = x_{k+1} - x$ is taken to be consistent with $\sinh(\kappa(a - r))$.

It is easy to prove that Lemma 9.2 is also valid for the proposed randomized approach to treating the continuity boundary conditions.

It is essential to note that a randomization of the FD approximation for the normal derivative with a step, h, provides an $O(h^2)$ bias, both for the Neumann and the continuity boundary conditions.

9.4.4 Construction of the algorithm and its convergence

To complete the construction, we need a Monte Carlo estimate for the solution value at an arbitrary point in the exterior domain, G_e. It is natural to use the estimate based on simulation of the WOS Markov chain. Every step of the algorithm is the direct randomization of the mean-value formula for the PB equation: $u(x) = \int_{S(x,d)} \frac{q(\kappa,d)}{4\pi d^2}\, u\, ds$, where $S(x,d)$ is the surface of a ball totally contained in G_e and d is usually taken to be the maximum possible, i.e., equal to the distance from x to Γ. For $\kappa = 0$, we use the modification of the WOS algorithm with the direct simulation of jump to the absorbing state of the Markov chain at infinity [24]. The conditional mean number of steps for the chain to hit the ε-strip near the boundary is $O(\log \varepsilon)$. For positive κ, we consider $q(\kappa, d)$ the survival probability. In this case, the WOS either comes to Γ or terminates at some finite absorbing state in G_e.

To prove the convergence of the algorithm, we reformulate the problem in terms of integral equations with generalized kernels [21, 23]. Inside the domains, G_i and G_e, we consider the mean-value formulas as such equations. In Γ_ε, we use the approximation $u(x) = u(x^*) + \phi(x, x^*)$ and substitute the integral representation for $u(x^*)$ at a boundary point, x^*. Hence, the described random-walk construction corresponds to the so-called direct simulation of (approximated) integral equation [23]. This means that the resulting estimate for the solution's value at a point, $x = x_0 \in \mathbb{R}^3$, is

$$\xi[u](x) = \sum_{i=0}^{N} \xi[F](x_i). \tag{9.32}$$

Here, N is the random length of the Markov chain, and $\xi[F]$ are estimates for the right-hand side of the integral equation. For the external Neumann problem, this function is

$$\begin{aligned} F(x) &= 0, \text{ when } x \in G_e \setminus \overline{\Gamma}_\varepsilon; \\ &= -\int\limits_{\Gamma \bigcap B(x^*,a)\setminus\{x^*\}} \frac{1}{2\pi r}\left(1 - \frac{r}{a}\right) Q^1_{\kappa,a}(r)\, f(y)\, ds(y) \\ &\quad + \phi_\Gamma + \phi(x, x^*), \text{ when } x \in \overline{\Gamma}_\varepsilon, \end{aligned} \tag{9.33}$$

where $Q^1_{\kappa,a}(r) = \frac{\sinh(\kappa(a-r))}{a-r} \frac{a}{\sinh(\kappa a)}$, $r = |y - x^*|$.

For Markov chains based on the direct simulation, finiteness of the mean number of steps, $\mathbb{E}N < \infty$, is equivalent to convergence of the Neumann series for the corresponding integral operator, which kernel coincides with the transition density of this Markov chain. Besides that, the kernel of the integral operator that defines the second moment of the estimate is also equal to this density [23]. This means that for the exactly known free term, F, estimate (9.32) is unbiased and has finite variance. The same is true if estimates, $\xi[F](x_i)$, are unbiased and have uniformly in x_i bounded second moments. It is clear that we can easily choose such a density that an estimate for the integral in (9.33) will have the requested properties.

To prove that the mean number of steps is finite, we consider the auxiliary BVP:

$$\Delta p_0(x) - \kappa^2 p_0(x) = 0, x \in G_e, \quad \frac{\partial p_0}{\partial n}|_\Gamma = -1. \tag{9.34}$$

For this problem, the integral in (9.33) equals $(\cosh(\kappa a) - 1)/(\kappa \sinh(\kappa a))$ when Γ is a plane or a sphere. This is equal to $\frac{a}{2}\left(1 - \frac{(\kappa a)^2}{24} + O(\kappa a)^4\right)$, as $\kappa a \to 0$. Setting $\varepsilon = O(a/2R)^3$, we have

Lemma 9.3. *The mean number of boundary hits in the WOS solving the exterior Neumann problem is* $\mathbb{E}N^* = \frac{2p_0}{a}\left(1 + O(a^2)\right)$.

In the full analogy, we obtain the following.

Lemma 9.4. *The mean number of boundary hits of the WOS algorithm solving the continuity BVP is* $\mathbb{E}N^* = \frac{2p_1}{a}(1 + O(a))$. *Here,* p_1 *is a bounded solution to the problem (9.24), (9.25) with no charges in* G_i *and with the boundary condition* $\epsilon_i \frac{\partial p_{1,i}}{\partial n}(y) = \epsilon_e \frac{\partial p_{1,e}}{\partial n}(y) + 1$.

Denote by $\{x^*_{k,j} \in \Gamma, j = 1, \dots, N^*_i\}$ the sequence of exit points from G_i for the WOS Markov chain used to calculate the solution of the continuity BVP. Let $\{x_{k+1,j} \in G_i, j = 1, \dots, N^*_i - 1\}$ be the sequence of return points. Clearly, $\mathbb{E}N^*_i = p_i \mathbb{E}\, N^*$. Then, we have

Theorem 9.5. *The quantity* $g(x_0) + \xi[u^0](x_0)$, *where*

$$\xi[u^0](x_0) = -g(x^*_{k,1}) + \sum_{j=1}^{N^*_i - 1} \left[g(x_{k+1,j}) - g(x^*_{k,j+1})\right] \tag{9.35}$$

is the estimate for the solution of the BVP (9.24), (9.25), (9.26).
For $\varepsilon = (a/2R)^3$, *the bias of this estimate is* $O(a)^2$ *as* $a \to 0$. *The variance of this estimate is finite, and the computational cost is* $O(\log(\delta)\, \delta^{-5/2})$, *for a given accuracy,* δ.

The variance is finite since the algorithm is based on the direct simulation of the transformed integral equation [23]. The logarithmic factor in the estimate comes from the mean number of steps in the WOS Markov chain until it hits for the first time the ε-strip near the boundary [21, 24].

The same simulation can be used for computing the gradient of the solution

$$\nabla u(x_0) = \nabla g(x_0) + \mathbb{E} Q_0 \xi[u^0](x_0).$$

The vector weight Q_0 is computed at the first step of the algorithm, when simulating the jump to the sphere $S(x_0, d(x_0))$ [21]. Let $x_0 \equiv (x_{0,(1)}, x_{0,(2)}, x_{0,(3)})$. Then,

$$\frac{\partial u^0}{\partial x_{0,(i)}}(x) = \mathbb{E}\, \frac{3(x_{1,(i)} - x_{0,(i)})}{d(x_0)^2}\, u_0(x_1),\ i = 1, 2, 3,$$

i.e., $Q_0 = 3(x_1 - x_0)/d(x_0)^2$.

It is essential to note that with the FD approximation of the normal derivative using the step h, the mean number of boundary hits is $O(h^{-1})$. Therefore, the bias of the resulting estimate is $O(h)$, and the computational cost is $O(\log(\delta)\ \delta^{-3})$ [57, 69]. Thus, even the simplest approximation to the (exact!) integral relation we constructed substantially improves efficiency of the WOS algorithm.

9.5 Computing macromolecule energy

Biological macromolecules, such as peptides and proteins, contain charged and polar groups and exist in a heterogeneous cellular environment, which contains ions, water and other small molecules. Thus, it is not surprising that in vivo and in vitro studies show that salt-mediated long-range electrostatic interactions play a crucial and often complex role in the structure, stability, folding, dynamics and binding behaviour of biomolecules. For instance, it is well known that the thermodynamic binding affinity of charged ligands (*e.g.* metal ions, proteins and peptides) to proteins, the kinetic rates of protein–protein association and the stability of proteins can change significantly with small changes in salt concentration of the ionic solution [1, 16, 33, 61, 66, 80, 117]. To calculate salt-mediated electrostatic interactions, all-atom explicit solvent molecular dynamics simulations of the biomolecules in their natural environment [27, 84, 119] could be used. However, due to the extreme complexity of such physical model that should involve an enormous number of atoms, this approach is not feasible for routine calculations. On the other hand, the linear PB equation, which is based on an implicit solvent model but still treats the biomolecule at the molecular level of detail, has been successfully used to predict and interpret numerous salt-mediated electrostatic effects [4, 17, 20, 48]. Performing linear PB calculations of salt-dependent biomolecular properties can be problematic due to the inherent approximations of the deterministic PB numerical methods normally used. Such approach includes discretization errors and a rather crude treatment of the boundary conditions. Here, we describe an application of a grid-free Monte Carlo linear PB solver that can deliver non-specific salt-dependent electrostatic properties such as electrostatic solvation free energies with very high accuracy.

9.5.1 Mathematical model and computational results

Within the framework of the adopted here implicit solvent model, a polarizable solvent (aqueous salt solution) is considered as an infinite dielectric continuum, in which a cavity G_i having the geometry of a molecule is immersed. Distribution of the electrostatic potential $\Phi(x)$ in the system satisfies the Poisson equation in the whole space (in CGS units):

$$-\nabla\epsilon\,\nabla\Phi(x)=4\pi\rho(x),\ x\in\mathbb{R}^3. \tag{9.36}$$

Inside the molecule, point charges $Q_m, m=1,\dots,M$ are located at the atomic centres, $x^{(m)}$, respectively. This means the potential $\Phi_i(x)$ satisfies

$$\epsilon_i\Delta\Phi_i(x)+4\pi\sum_{m=1}^{M}Q_m\delta(x-x^{(m)})=0,\ x\in G_i, \tag{9.37}$$

where ϵ_i is constant dielectric permittivity.

Outside the molecule, the charge distribution is determined by the mobile ions. In the equilibrium state, concentration of a particular kind of ions, j, is described by the Boltzmann law:

$$M_j(x)=M_j^0\exp\left(\frac{-z_je\Phi(x)}{k_BT}\right).$$

Here, z_je is the charge of a solitary ion, e is the protonic charge, k_B is the Boltzmann's constant and T is the absolute temperature. In supposition that there is only 1:1 salt in the solution that dissociates into an equal number of ions of opposite sign, we get

$$\rho(x)=-2ez_1M\sinh\left(\frac{z_1e\Phi(x)}{k_BT}\right),$$

where $z_1=-z_2=1$, $M_1^0=M_2^0\equiv M$. Denote by $\epsilon_e\geq\epsilon_i$ the dielectric permittivity of the solution. Then,

$$\epsilon_e\Delta\Phi_e(x)-8\pi ez_1M\sinh\left(\frac{z_1e\Phi_e(x)}{k_BT}\right)=0,\ x\in G_e, \tag{9.38}$$

Here, $\Phi_e(x)$ is the potential outside the molecule, $G_e=\mathbb{R}^3\setminus\overline{G_i}$. Denote $c_j=1000M_j^0/N_A$ the molar concentration, where N_A is the Avogadro number, and introduce the ionic strength of solution (in M),

$$I_S=\frac{1}{2}\sum_{j=1}^{N_I}c_jz_j^2.$$

In this particular case, $N_I=2$ and $I_S=1000M/N_A$.

From here, we get that the dimensionless (unitless) electrostatic potential $u = \frac{e}{k_B T}\Phi$ satisfies the non-linear PB equation

$$\Delta u_e(x) - \kappa^2 \sinh(u_e(x)) = 0, \ x \in G_e. \tag{9.39}$$

Here,

$$\kappa^2 = \frac{8\pi N_A e^2 I_s}{\epsilon_e 1000 k_B T} \tag{9.40}$$

is the square of the Debye-Hückel parameter (or inverse length).

For small salt concentration and values of the electrostatic potential, equation (9.39) may be linearized. In the linear case, we can get back to $\Phi_e(x)$, which satisfies the following equation:

$$\Delta\Phi_e(x) - \kappa^2\Phi_e(x) = 0, \ x \in G_e. \tag{9.41}$$

The system of equations (9.37), (9.41) must be complemented by the boundary conditions at infinity: $\Phi_e(x) \to 0$ as $|x| \to \infty$ and on the boundary between G_i and G_e. On that surface, the continuity conditions for both potential and flux must be satisfied [41, 113]:

$$\Phi_i = \Phi_e, \quad \epsilon_i \frac{\partial\Phi_i}{\partial n} = \epsilon_e \frac{\partial\Phi_e}{\partial n}. \tag{9.42}$$

In the simplest model considered here, the dielectric boundary coincides with the van der Waals surface (see Fig. 9.2). However, the algorithms, which are described here, can be used for more complicated models as well, including the model with the ion-exclusion layer [10, 118].

One of the most important macroparameters for a molecule is its free electrostatic energy. In the linear case, it can be represented as a linear functional of the potential point values:

$$W_{rf} = \frac{1}{2}\sum_{m=1}^{M} Q_m \Phi_{rf}(x^{(m)}). \tag{9.43}$$

Here, Φ_{rf} is the electrostatic potential, which is due to the reaction of electric field on the dielectric boundary and charges outside the molecule. Note that its values have to be computed exactly at the positions of point charges.

Remind that the original problem (9.36) is stated in CGS units. At the atomic scale, however, it is natural to measure distances in angstroms (Å) and charges in proton (e) units. In these units, κ is of order 1 Å^{-1}, while in CGS units, κ is of order 10^8 cm^{-1}. Therefore, all computations are performed in atomic units. After that, the calculated value of energy in units $e^2/\text{Å}$ is multiplied by 332.06364 to get the result measured in kcal/mol.

With the Monte Carlo algorithms, it is possible to separate singularities and to compute the reaction field only. To do this, we use the following representation:

$$\Phi_i(x) = \Phi_{rf}(x) + \Phi^c(x), \tag{9.44}$$

where the singular part can be written down explicitly:

$$\Phi^c(x) = \sum_{m=1}^{M} \frac{Q_m}{\epsilon_i |x - x^{(m)}|}. \tag{9.45}$$

To compute the reaction field potential, we use estimate (9.35), where $g = \Phi^c$.

Consider now the problem of estimating the dependence of energy on salt concentration in the solution. For linear model and 1:1 salt, this dependence manifests itself through the Debye-Hückel parameter, κ. Let $0 < \kappa_1 < \kappa_2 < \ldots < \kappa_{n\kappa}$ be the parameter values, for which the energy should be estimated. Clearly, the survival probability, $q(\kappa, d)$ (see (9.16)), for the random walk in the outside domain, G_e, monotonically decreases when κ increases. Therefore, the minimal value (κ_1) is used for simulation, while energy estimates for other values of this parameter are calculated using the weight functions, $F_j(\kappa)$:

$$\xi[\Phi_{rf}](x^{(m)}) = -\Phi^c(x_1^*) + \sum_{j=2}^{N_{ins}} F_j(\kappa) \left(\Phi^c(x_j^{ins}) - \Phi^c(x_{j,ins}^*)\right), \tag{9.46}$$

where $F_j(\kappa) \equiv 1$ for $\kappa = \kappa_1$, and at every step in G_e, it is multiplied by $q(\kappa_i, d_k)/q(\kappa_1, d_k)$ for all other values of κ. Thus, we have

$$\xi[W_{rf}] = \frac{1}{2} \sum_{m=1}^{M} Q_m \xi[\Phi_{rf}](x^{(m)}), \tag{9.47}$$

where salt-dependent estimates for the reaction field potential are given by (9.46).

This estimate can be used for computing the electrostatic solvation free energy (ΔG_{elec}^{solv}). To do this, we need to calculate the difference between two results: the first one for the current values of ϵ_i, ϵ_e and κ and the second (reference) one for $\epsilon_e = 1$, $\kappa = 0$, i.e., for the molecule in vacuum.

In Figure 9.3, we present the results of computations for the particular macromolecule, namely, calmodulin (3cln in the publicly available protein database) in comparison with the FD results. These results fall well within the 95% confidence interval of the Monte Carlo (MC) energy values at all salt concentrations. Moreover, in the limit of zero salt concentration, the salt derivative of ΔG_{elec}^{solv} is equal to zero for both MC- and FD-based PB codes. This property coincides with the theoretically predicted behaviour.

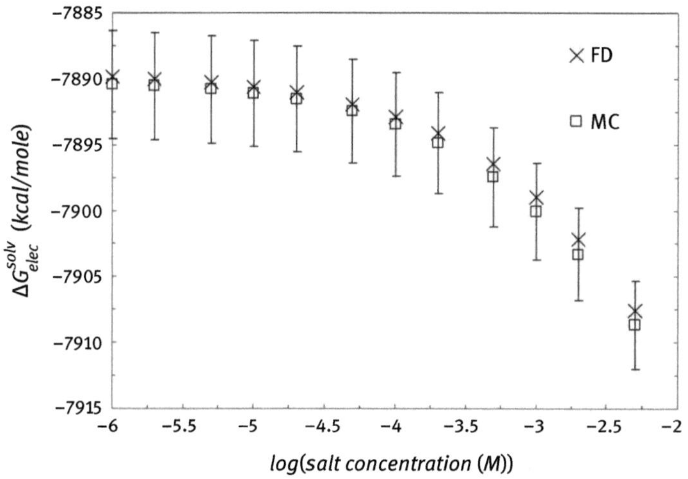

Fig. 9.3. Comparison of the salt dependence of the electrostatic solvation free energy (ΔG_{elec}^{solv}) of calmodulin (PDB id: 3cln) obtained with two independent Poisson-Boltzmann (PB) solvers: Monte Carlo estimation (MC) and finite-difference-based software (FD).

It is essential to note that every point of the deterministic curve required a separate PB computation, whereas the Monte Carlo algorithm made it possible to calculate the curve in a single computation.

Bibliography

[1] I. Andre, T. Kesvatera, B. Jonsson and S. Linse, Salt enhances calmodulin-target interaction, *Biophys. J.* **90** (2006), 2903–2910.

[2] S. Arya and D. M. Mount, Approximate nearest neighbor queries in fixed dimensions, in: *Proc. 4th ACM-SIAM Symp. Discrete Algorithms*, pp. 271–280, 1993.

[3] S. Arya, D. M. Mount, N. S. Netanyahu, R. Silverman and A. Wu, An optimal algorithm for approximate nearest neighbor searching, *J. ACM* **45** (1998), 891–923.

[4] P. Barth, T. Alber and P. B. Harbury, Accurate, conformation-dependent predictions of solvent effects on protein ionization constants, *Proc. Natl Acad. Sci. USA* **104** (2007), 4898–4903.

[5] K. Binder, K. Binder, D. Ceperley, J. Hansen, M. Kalos, D. Landau, D. Levesque, H. Müller-Krumbhaar, D. Stauffer and J. Weis, *Monte Carlo Methods in Statistical Physics*, Topics in Current Physics, Springer, Berlin, Heidelberg, 2012.

[6] S. Boggs and D. Krinsley, *Application of cathodoluminescence imaging to the study of sedimentary rocks*, Cambridge University Press, 2006.

[7] M. Bramson and J. L. Lebowitz, Asymptotic behavior of densities for two-particle annihilating random walks, *J. Stat. Phys.* **62** (1991), 297–372.

[8] O. Brandt and K. H. Ploog, Solid state sighting: The benefit of disorder, *Nat. Mater.* **5** (2006), 769–770.

[9] M. A. Caro, S. Schulz and E. P. O'Reilly, Theory of local electric polarization and its relation to internal strain: Impact on polarization potential and electronic properties of group-III nitrides, *Phys. Rev. B* **88** (2013), 214103.

[10] M. L. Connolly, The molecular surface package, *J. Mol. Graph.* **11** (1993), 139–141.

[11] R. W. Conway, Some tactical problems in digital simulation, *Manage. Sci.* **10** (1963), 47–61.

[12] R. Courant and D. Hilbert, *Methods of Mathematical Physics*, **2**, John Wiley & Sons, New York, 1989.

[13] D. R. Cox, *Renewal Theory*, Methuen, London, 1962.

[14] M. E. Davis and J. A. McCammon, Electrostatics in biomolecular structure and dynamics, *Chem. Rev.* **90** (1990), 509–521.

[15] M. de Berg, O. Cheong, M. van Kreveld and M. Overmars, *Computational Geometry: Algorithms and Applications*, 3rd revised ed, Springer-Verlag, 2008.

[16] M. A. de los Rios and K. W. Plaxco, Apparent Debye–Hückel electrostatic effects in the folding of a simple, single domain protein, *Biochemistry* **44** (2005), 1243–1250.

[17] B. N. Dominy, D. Perl, F. X. Schmid and C. L. Brooks III, The effects of ionic strength on protein stability: The cold shock protein family, *J. Mol. Biol.* **319** (2002), 541–554.

[18] A. Donev, V. V. Bulatov, T. Oppelstrup, G. H. Gilmer, B. Sadigh and M. H. Kalos, A First-passage kinetic Monte Carlo algorithm for complex diffusion-reaction systems, *J. Comput. Phys.* **229** (2010), 3214–3236.

[19] E. B. Dynkin, *Markov Processes*, Fizmatgiz, Moscow, 1963 (in Russian).

[20] A. H. Elcock and J. A. McCammon, Electrostatic contributions to the stability of halophilic proteins, *J. Mol. Biol.* **280** (1998), 731–748.

[21] B. S. Elepov, A. A. Kronberg, G. A. Mikhailov and K. K. Sabelfeld, *Solution of Boundary Value Problems by the Monte Carlo Method*, Nauka, Novosibirsk, 1980 (in Russian).

[22] D. L. Ermak and J. A. McCammon, Brownian dynamics with hydrodynamic interactions, *J. Chem. Phys.* **69** (1978), 1352–1360.

[23] S. M. Ermakov and G. A. Mikhailov, *Statistical simulation*, Nauka, Moscow, 1982 (in Russian).

[24] S. M. Ermakov, V. V. Nekrutkin and A. S. Sipin, *Random processes for classical equations of mathematical physics*, Kluwer Academic Publishers, Dodrecht, The Netherlands, 1989.

[25] D. Faddeev and V. Faddeeva, *Numerical Methods of Linear Algebra*, Fizmatgiz, Moscow, 1963 (in Russian).
[26] W. Feller, *An Introduction to Probability Theory and Its Applications*, **2**, Wiley, New York, 1968.
[27] M. S. Formaneck, L. Ma and Q. Cui, Effects of temperature and salt concentration on the structural stability of human lymphotactin: Insights from molecular simulations, *J. Am. Chem. Soc.* **128** (2006), 9506–9517.
[28] M. Freidlin, *Functional Integration and Partial Differential Equations*, Princeton University Press, Princeton, 1985.
[29] A. Friedman, *Stochastic Differential Equations and Applications*, **1, 2**, Academic Press, New York, 1976.
[30] J. A. Given, J. B. Hubbard and J. F. Douglas, A first passage algorithm for the hydrodynamic friction and diffusion-limited reaction rate of macromolecules, *J. Chem. Phys.* **106** (1997), 3761–3771.
[31] N. Golyandina, Convergence rate for spherical processes with shifted centres, *Monte Carlo Methods and Applications* **10** (2004), 287–296.
[32] E. Goursat, *Cours D'Analyse Mathematique III*, Gauthier-Villars, Paris, 1930 (in French).
[33] R. A. Grucza, J. M. Bradshaw, V. Mitaxov and G. Waksman, Role of electrostatic interactions in SH2 domain recognition: Salt-dependent of tyrosyl-phosphorylated peptide binding to the tandem SH2 domain of the Syk kinase and the single SH2 domain of the Src kinase, *Biochemistry* **39** (2000), 10072–10081.
[34] N. M. Günter, *La theorie du potentiel et ses applications aux problemes fondamentaux de la physique mathematique*, Gauthier-Villars, Paris, 1934 (in French).
[35] A. Haji-Sheikh and E. M. Sparrow, The floating random walk and its application to Monte Carlo solutions of heat equations, *SIAM J. Appl. Math.* **14** (1966), 570–589.
[36] J. M. Hammersley and D. C. Handscomb, *Monte Carlo Methods*, Methuen, London, 1964.
[37] S. Hammersley, D. Watson-Parris, P. Dawson, M. J. Godfrey, T. J. Badcock, M. J. Kappers, C. McAleese, R. A. Oliver and C. J. Humphreys, The consequences of high injected carrier densities on carrier localization and efficiency droop in InGaN/GaN quantum well structures, *J. Appl. Phys.* **111** (2012).
[38] H. Hennion and L. Herve, *Limit theorems for Markov Chains and Stochastic Properties of Dynamical Systems by Quasi-Compactness*, Springer-Verlag, Berlin, Heidelberg, 2001.
[39] A. Iljin, A. Kalashnikov and O. Olejnik, Second order parabolic linear equations, *Usp. Mat. Nauk.* **17** (1962), 3–147 (in Russian).
[40] P. Jäckel, *Monte Carlo Methods in Finance*, John Wiley & Sons, New York, 2002.
[41] J. D. Jackson, *Classical Electrodynamics*, John Wiley & Sons, New York, 1962.
[42] K. Jörgens, *Linear Integral Operators*, Surveys and Reference Works in Mathematics, Pitman Advanced Publishing Program, Boston London Melbourne, 1982.
[43] M. Kac, On some connection between probability theory and differential and integral equations, in: *Proc. of the Second Berkeley Symp. on Mathematical Statistics and Probability*, University of California Press, pp. 189–215, 1951.
[44] G. Kallianpur, *Stochastic Filtering Theory*, Stochastic Modelling and Applied Probability, Springer, New York, 2013.
[45] V. A. Kanevsky and G. S. Lev, On simulation of the exit point for a Brownian motion in a ball, *Russ. J. Appl. Math. Phys.* **17** (1977), 763–764 (in Russian).
[46] L. W. Kantorovich and V. I. Krylov, *Approximate Methods of Higher Analysis*, Interscience, New York, 1964.
[47] L. W. Kantorowitsch and G. P. Akilow, *Funktionalanalysis in Normierten Räumen*, Akademie Verlag, Berlin, 1964 (in German).

[48] Y.-H. Kao, C. A. Fitch and E. B. Garcia-Moreno, Salt effects on ionization equilibria of histidines in myoglobin, *Biophys. J.* **79** (2000), 1637–1654.
[49] A. Karaivanova, M. Mascagni and N. A. Simonov, Parallel quasi-random walks on the boundary, *Monte Carlo Methods Appl.* **10** (2004), 311–320.
[50] A. Karaivanova, M. Mascagni and N. A. Simonov, Solving BVPs using quasirandom walks on the boundary, *Lecture Notes Comput. Sci.* **2907** (2004), 162–169.
[51] A. E. Kireeva and K. K. Sabelfeld, Cellular automata model of electrons and holes annihilation in an inhomogeneous semiconductor, *Lecture Notes Comput. Sci.* **9251** (2015), 191–200.
[52] P. E. Kloeden and E. Platen, *Numerical Solution of Stochastic Differential Equations*, Springer-Verlag, 1992.
[53] A. Kolodko, K. Sabelfeld and W. Wagner, A stochastic method for solving Smoluchowski's coagulation equation, *Math. Comput. Simul.* **49** (1999), 57–79.
[54] A. A. Kolodko and K. K. Sabelfeld, Stochastic Lagrangian model for spatially inhomogeneous Smoluchowski equation governing coagulating and diffusing particles, *Monte Carlo Methods Appl.* **7** (2001), 223–228.
[55] V. S. Korolyuk, N. I. Portenko, A. V. Skorokhod and A. F. Turbin, *Handbook of Probability Theory and Mathematical Statistics*, Nauka, Moscow, 1985 (in Russian).
[56] E. Kotomin and V. Kuzovkov (eds.), *Modern Aspects of Diffusion-Controlled Reactions. Cooperative Phenomena in Bimolecular Processes*, Comprehensive Chemical Kinetics **34**, Elsevier, Amsterdam, 1996.
[57] A. Kronberg, On algorithms for statistical simulation of the solution of boundary value problems of elliptic type, *Zh. Vychisl. Mat. i Mat. Phyz.* **84** (1984), 1531–1537 (in Russian).
[58] V. D. Kupradze, T. G. Gegelia, M. O. Basheleishvili and T. V. Burchuladze, *Three-Dimensional Problems of Mathematical Theory of Elasticity and Thermoelasticity*, North Holland, Amsterdam, 1979.
[59] O. Kurbanmuradov, K. K. Sabelfeld and N. A. Simonov, *Random walk on Boundary Algorithms*, Computing Center SB USSR Academy of Sciences, Novosibirsk, 1989 (in Russian).
[60] O. A. Ladyzhenskaja, V. A. Solonnikov and N. N. Uraltseva, *Linear and Quasilinear Equations of Parabolic Type*, Nauka, Moscow, 1967 (in Russian).
[61] J. Lanyi, Salt-dependent properties from extremely halophilic bacteria, *Bacteriol. Rev.* **38** (1974), 272–290.
[62] B. A. Luty, J. A. McCammon and H.-X. Zhou, Diffusive reaction rates from Brownian dynamics simulations: Replacing the outer cutoff surface by an analytical treatment, *J. Chem. Phys.* **97** (1992), 5682–5686.
[63] I. Marek, On the approximative construction of the eigenvectors corresponding to a pair of complex conjugated eigenvalues, *Mat.-Fyzikalny Cas. SAV* **14** (1964), 277–288.
[64] M. Mascagni and N. A. Simonov, Monte Carlo methods for calculating some physical properties of large molecules, *SIAM J. Sci. Comput.* **26** (2004), 339–357.
[65] J. C. Maxwell, *A Treatise on Electricity and Magnetism*, 3rd ed, **1**, Oxford University Press, 1892.
[66] R. H. Meltzer, E. Thompson, K. V. Soman, X.-Z. Song, J. O. Ebalunode, T. G. Wensel, J. M. Briggs and S. E. Pedersen, Electrostatic steering at acetylcholine binding sites, *Biophys. J.* **91** (2006), 1302–1314.
[67] N. Metropolis, A. W. Rosenbluth, M. N. Rosenbluth, A. H. Teller and E. Teller, Equation of state calculations by fast computing machines, *J. Chem. Phys.* **21** (1953), 1087–1092.
[68] S. P. Meyn and R. L. Tweedie, *Markov Chains and Stochastic Stability*, Springer-Verlag, London, 1993.

[69] G. A. Mikhailov and R. N. Makarov, Solution of boundary value problems by the "random walk on spheres" method with reflection from the boundary, *Dokl. Akad. Nauk* **353** (1997), 720–722 (in Russian).
[70] S. G. Mikhlin, *Multidimensional Singular Integrals and Integral Equations*, Fizmatgiz, Leningrad, 1962 (in Russian).
[71] G. N. Milstein, *Numerical Integration of Stochastic Differential Equations*, Kluwer Academic Publishers, Dordrecht, The Netherlands, 1994.
[72] C. Miranda, *Partial Differential Equations of Elliptic Type*, Springer-Verlag, Berlin, Heidelberg, New York, 1970.
[73] B. A. Mishustin, On the solution of the Dirichlet problem for Laplace equation by a statistical testing method, *Zhurn. Vychisl. Matem. i Matem. Fiz.* **7** (1967), 1178–1187 (in Russian).
[74] S. Mizohata, *The Theory of Partial Differential Equations*, Cambridge University Press, Cambridge, 1973.
[75] M. Motoo, Some evaluations for continuous Monte Carlo method by using Brownian hitting process, *Ann. Inst. Stat. Math.* **11** (1959), 49–55.
[76] M. E. Müller, Some continuous Monte Carlo methods for the Dirichlet problem, *Ann. Math. Stat.* **27** (1956), 569–589.
[77] D. Nakaji, V. Grillo, N. Yamamoto and T. Mukai, Contrast analysis of dislocation images in TEM-cathodoluminescence technique, *J. Electron Microsc.* **54** (2005), 223–230.
[78] S. H. Northrup, S. A. Allison and J. A. McCammon, Brownian dynamics simulation of diffusion-influenced bimolecular reactions, *J. Chem. Phys.* **80** (1984), 1517–1524.
[79] T. Opplestrup, V. V. Bulatov, G. H. Gilmer, M. H. Kalos and B. Sadigh, First-passage Monte Carlo algorithm: Diffusion without all the hops, *Phys. Rev. Lett.* **97** (2006), 230602.
[80] A. Ortiz-Baerga, A. R. Rezaie and E. A. Komives, Electrostatic dependence of the thrombin-thromomodulin interaction, *J. Mol. Biol.* **296** (2000), 651–658.
[81] A. A. Ovchinnikov and Y. B. Zeldovich, Role of density fluctuations in bimolecular reaction kinetics, *Chem. Phys.* **28** (1978), 215 – 218.
[82] K. L. Pey, D. S. H. Chan, J. C. H. Phang, J. F. Breese and S. Myhajlenko, Cathodoluminescence contrast of localized defects part I. Numerical model for simulation, *Scanning Microsc.* **9** (1995), 355–366.
[83] G. Polya and G. Szego, *Aufgaben und Lehrsätze aus der Analysis*, Springer, Berlin, Heidelberg, 1964 (in German).
[84] S. Y. Ponomarev, K. M. Thayer and D. L. Beveridge, Ion motions in molecular dynamics simulations on DNA, *Proc. Natl Acad. Sci. USA* **101** (2004), 14771–14775.
[85] A. P. Prudnikov, J. F. Brychkov and O. I. Marichev, *Integrals and Series*, Nauka, Moscow, 1981 (in Russian).
[86] A. P. Prudnikov, J. F. Brychkov and O. I. Marichev, *Integrals and Series: Special Functions*, Nauka, Moscow, 1983 (in Russian).
[87] A. G. Ramm, *Iterative Methods for Calculating Static Fields and Wave Scattering by Bodies*, Springer, New York, 1982.
[88] C. P. Rao, *Linear Statistical Inference and Its Applications*, John Wiley and Sons, New York, London, Sydney, 1960.
[89] M. L. Rasulov, *Application of Contour Integrals to Solving Problems for Parabolic Systems of Second Order*, Nauka, Moscow, 1975 (in Russian).
[90] S. Redner, *A Guide to First-passage Processes*, Cambridge University Press, Cambridge, 2001.
[91] S. A. Rice, *Diffusion-Limited Reactions*, Comprehensive Chemical Kinetics **25**, Elsevier, Amsterdam, The Netherlands, 1985.
[92] R. Y. Rubinstein and D. P. Kroese, *Simulation and the Monte Carlo Method*, 2 ed, Wiley, New York, 2008.

[93] K. Sabelfeld, A. Levykin and A. Kireeva, Stochastic simulation of fluctuation-induced reaction-diffusion kinetics governed by Smoluchowski equations, *Monte Carlo Methods Appl.* **21** (2015), 33–48.
[94] K. Sabelfeld and N. Mozartova, Sparsified randomization algorithms for large systems of linear equations and a new version of the random walk on boundary method, *Monte Carlo Methods Appl.* **15** (2009), 257–284.
[95] K. K. Sabelfeld, Vector Monte Carlo algorithms for solving second order elliptic systems and the Lame equation, *Dokl. Akad. Nauk SSSR* **262** (1982), 1076–1080 (in Russian).
[96] K. K. Sabelfeld, The walk along the boundary algorithm for solving boundary value problems, *Sov. J. Num. Anal. Math. Modelling* **1** (1986), 18–28.
[97] K. K. Sabelfeld, The ergodic walk on boundary process for solving Robin problems, In "Theory and applications of statistical modelling", Novosibirsk, Computing Center, *Russ. Acad. Science*, 1989, 118–123 (in Russian)
[98] K. K. Sabelfeld, *Monte Carlo Methods in Boundary Value Problems*, Springer-Verlag, Berlin, Heidelberg, New York, 1991.
[99] K. K. Sabelfeld, *Stochastic Algorithms for Solving Smolouchovsky Coagulation Equation*, Stochastic Simulation (S. Ogawa and K. Sabelfeld, eds.), Kyoto, 1997, pp. 80–105.
[100] K. K. Sabelfeld, O. Brandt and V. M. Kaganer, Stochastic model for the fluctuation-limited reaction-diffusion kinetics in inhomogeneous media based on the nonlinear Smoluchowski equations, *J. Math. Chem.* **53** (2015), 651–669.
[101] K. K. Sabelfeld and A. A. Kolodko, Stochastic Lagrangian models and algorithms for spatially inhomogeneous Smoluchowski equation, *Math. Comput. Simul.* **61** (2003), 115–137.
[102] K. K. Sabelfeld, S. V. Rogasinsky, A. A. Kolodko and A. Levykin, Stochastic algorithms for solving Smolouchovsky coagulation equation and applications to aerosol growth simulation, *Monte Carlo Methods Appl.* **2** (1996), 41–87.
[103] K. K. Sabelfeld and N. A. Simonov, Walk along the boundary algorithms for solving boundary value problems, *Chisl. Metody Mech. Splosh. Sredy* **14** (1983), 116–134 (in Russian).
[104] K. K. Sabelfeld and N. A. Simonov, *Random Walks on Boundary for solving PDEs*, VSP, Utrecht, The Netherlands, 1994.
[105] D. Shoup, G. Lipari and A. Szabo, Diffusion-controlled bimolecular reaction rates, *Biophys. J.* **36** (1981), 697–714.
[106] N. A. Simonov, *A Random Walk Algorithm for Solving Boundary Value Problems with Partition into Subdomains*, Metody i algoritmy statisticheskogo modelirovanija [Methods and algorithms for statistical modelling], Akad. Nauk SSSR Sibirsk. Otdel., Vychisl. Tsentr, Novosibirsk, pp. 48–58, 1983. (in Russian).
[107] N. A. Simonov, Monte Carlo methods for solving elliptic equations with boundary conditions containing the normal derivative, *Dokl. Math.* **74** (2006), 656–659.
[108] N. A. Simonov, Random walk on spheres algorithms for solving mixed and Neumann boundary value problems, *Siberian J. Numer. Math.* **10** (2007), 209–220 (in Russian).
[109] N. A. Simonov, Random walks for solving boundary-value problems with flux conditions, *Lecture Notes Comput. Sci.* **4310** (2007), 181–188.
[110] N. A. Simonov, Walk-on-spheres algorithm for solving boundary-value problems with continuity flux conditions, in: *Monte Carlo and Quasi-Monte Carlo Methods 2006* (A. Keller, S. Heinrich and H. Niederreiter, eds.), Springer-Verlag, Heidelberg, pp. 633–644, 2008.
[111] W. I. Smirnov, *Lehrgang der Höheren Mathematik*, **IV** and **V**, Berlin, 1962 (in German).
[112] M. Smoluchowski, Drei Vorträge über Diffusion, Brownische Molekularbewegung und Koagulation von Kolloidteilchen, *Phys. Z.* **17** (1916), 557–571, *ibid.* 585–599.
[113] W. R. Smythe, *Static and Dynamic Electricity*, 3rd ed, McGraw-Hill, New York, 1989.
[114] I. M. Sobol, *Numerical Monte Carlo Methods*, Nauka, Moscow, 1973 (in Russian).

[115] K. Solc and W. H. Stockmayer, Kinetics of diffusion-controlled reaction between chemically asymmetric molecules. I. General theory, *J. Chem. Phys.* **54** (1971), 2981–2988.

[116] S. Steisunas, On the sojourn time of the Brownian process in a multidimensional sphere, *Nonlinear Analysis: Modelling and Control* **14** (2009), 389–396.

[117] B. Svensson, B. Jonsson, E. Thulin and C. E. Woodward, Binding of Ca^{2+} to Calmodulin and its tryptic fragments: Theory and experiment, *Biochemistry* **32** (1993), 2828–2834.

[118] J. M. J. Swanson, J. Morgan and J. A. McCammon, Limitations of atom-centered dielectric functions in implicit solvent models, *J. Phys. Chem. B* **109** (2005), 14769–14772.

[119] A. S. Thomas and A. H. Elcock, Direct observation of salt effects on molecular interactions through explict-solvent molecular dynamics simulations: Differential effects on electrostatic and hydrophobic interactions and comparisons to Poisson–Boltzmann theory, *J. Am. Chem. Soc.* **128** (2006), 7796–7806.

[120] V. S. Vladimirov, *Equations of Mathematical Physics*, 4th, corr. ed, Nauka, Moscow, 1981 (in Russian).

[121] W. Wagner, Estimation by the Monte Carlo method of the generalized integrals of the principal value type, *Zhurn. Vychisl. Matem. i. Matem. Phys.* **22** (1982), 573–581 (in Russian).

[122] H.-X. Zhou, Comparison of three Brownian-dynamics algorithms for calculating rate constants of diffusion-influenced reactions, *J. Chem. Phys.* **108** (1998), 8139–8145.

[123] H.-X. Zhou, Theory of the diffusion-influenced substrate binding rate to a buried and gated active site, *J. Chem. Phys.* **108** (1998), 8146–8154.

[124] H.-X. Zhou, A. Szabo, J. F. Douglas and J. B. Hubbard, A Brownian dynamics algorithm for calculating the hydrodynamics friction and the electrostatic capacitance of an arbitrary shaped object, *J. Chem. Phys.* **100** (1994), 3821–3826.

Bei Fragen zur Produktsicherheit wenden Sie sich bitte an:
If you have any questions regarding product safety,
please contact:

Walter de Gruyter GmbH
Genthiner Straße 13
10785 Berlin
productsafety@degruyterbrill.com